Advanced Concept Maps in STEM Education:

Emerging Research and Opportunities

Michael Tang
University of Colorado Denver, USA

Arunprakash T. Karunanithi
University of Colorado Denver, USA

A volume in the Advances in Early Childhood and K-12 Education (AECKE) Book Series

IGI Global
DISSEMINATOR OF KNOWLEDGE

Published in the United States of America by
IGI Global
Information Science Reference (an imprint of IGI Global)
701 E. Chocolate Avenue
Hershey PA, USA 17033
Tel: 717-533-8845
Fax: 717-533-8661
E-mail: cust@igi-global.com
Web site: http://www.igi-global.com

Library of Congress Cataloging-in-Publication Data

Names: Tang, Michael, 1943- | Karunanithi, Arunprakash T., author.
Title: Advanced concept maps in STEM education: emerging research and opportunities / by Michael Tang and Arunprakash T. Karunanithi.
Description: Hershey PA : Information Science Reference, [2018]
Identifiers: LCCN 2016056029| ISBN 9781522521846 (hardcover) | ISBN 9781522521853 (ebook)
Subjects: LCSH: Science--Study and teaching (Elementary) | Science--Study and teaching (Middle school) | Concept mapping. | Geographic information systems.
Classification: LCC LB1585 .T28 2017 | DDC 372.35/044--dc23 LC record available at https://lccn.loc.gov/2016056029

This book is published in the IGI Global book series Advances in Early Childhood and K-12 Education (AECKE) (ISSN: 2329-5929; eISSN: 2329-5937)

British Cataloguing in Publication Data
A Cataloguing in Publication record for this book is available from the British Library.

Advances in Early Childhood and K-12 Education (AECKE) Book Series

ISSN:2329-5929
EISSN:2329-5937

Editor-in-Chief: Jared Keengwe, University of North Dakota, USA

MISSION

Early childhood and K-12 education is always evolving as new methods and tools are developed through which to shape the minds of today's youth. Globally, educational approaches vary allowing for new discussions on the best methods to not only educate, but also measure and analyze the learning process as well as an individual's intellectual development. New research in these fields is necessary to improve the current state of education and ensure that future generations are presented with quality learning opportunities.

The Advances in Early Childhood and K-12 Education (AECKE) series aims to present the latest research on trends, pedagogies, tools, and methodologies regarding all facets of early childhood and K-12 education.

COVERAGE

- Head Start and Pre-K Programs
- STEM Education
- Bullying in the Classroom
- Individualized Education
- Urban K-12 Education
- K-12 Education
- Curriculum Development
- Pedagogy
- Diverse Learners
- Poverty and Education

IGI Global is currently accepting manuscripts for publication within this series. To submit a proposal for a volume in this series, please contact our Acquisition Editors at Acquisitions@igi-global.com or visit: http://www.igi-global.com/publish/.

The Advances in Early Childhood and K-12 Education (AECKE) Book Series (ISSN 2329-5929) is published by IGI Global, 701 E. Chocolate Avenue, Hershey, PA 17033-1240, USA, www.igi-global.com. This series is composed of titles available for purchase individually; each title is edited to be contextually exclusive from any other title within the series. For pricing and ordering information please visit http://www.igi-global.com/book-series/advances-early-childhood-education/76699. Postmaster: Send all address changes to above address.

Titles in this Series

For a list of additional titles in this series, please visit:
http://www.igi-global.com/book-series/advances-early-childhood-education

Challenges Associated with Cross-Cultural and At-Risk Student Engagement
Richard K. Gordon (California State University, USA) Taichi Akutsu (Seisa University, Japan & Shujitsu University, Japan) J. Cynthia McDermott (Antioch University, USA) and Jose W. Lalas (University of Redlnds, USA)
Information Science Reference • ©2017 • 297pp • H/C (ISBN: 9781522518945) • US $180.00

Supporting the Education of Children with Autism Spectru Disorders
Yefim Kats (Chestnut Hill College, USA)
Information Science Reference • ©2017 • 336pp • H/C (ISBN: 9781522508168) • US $185.00

Optimizing K-12 Education through Online and Blended Learning
Nathaniel Ostashewski (Athabasca University, Canada) Jennifer Howell (Curtin University, Australia) and Martha Cleveland-Innes (Athabasca University, Canada)
Information Science Reference • ©2017 • 292pp • H/C (ISBN: 9781522505075) • US $185.00

Revolutionizing K-12 Blended Learning through the i²Flex Classroom Model
Maria D. Avgerinou (American Community Schools (ACS) Athens, Greece) and Stefanos P. Gialamas (American Community Schools (ACS) Athens, Greece)
Information Science Reference • ©2016 • 477pp • H/C (ISBN: 9781522502678) • US $175.00

Critical Examinations of School Violence and Disturbance in K-12 Education
Gordon A. Crews (Tiffin University, USA)
Information Science Reference • ©2016 • 322pp • H/C (ISBN: 9781466699359) • US $185.00

Improving K-12 STEM Education Outcomes through Technological Integration
Michael J. Urban (Bemidji State University, USA) and David A. Falvo (Walden University, USA)
Information Science Reference • ©2016 • 498pp • H/C (ISBN: 9781466696167) • US $215.00

Challenges Surrounding the Education of Children with Chronic Diseases
Maria Gordon (Voyageur Memorial Elementary School, Canada)
Information Science Reference • ©2016 • 402pp • H/C (ISBN: 9781466694521) • US $185.00

For an enitre list of titles in this series, please visit:
http://www.igi-global.com/book-series/advances-early-childhood-education

IGI Global
DISSEMINATOR OF KNOWLEDGE

701 East Chocolate Avenue, Hershey, PA 17033, USA
Tel: 717-533-8845 x100 • Fax: 717-533-8661
E-Mail: cust@igi-global.com • www.igi-global.com

Table of Contents

Preface

Strictly speaking, the term "concept maps" should be reserved to the definition J.D. Novak coined in the 1970s when he and his research team at Cornell developed the technique of concept mapping as a means of representing the emerging science knowledge of students with whom they were working. Since then concept mapping, which has its modern origin in the constructivist education movement of which Novak was a leading proponent, has been used as a tool to increase meaningful learning in the sciences and other subjects as well as to represent expert knowledge in government and business. Educational constructivists maintain that effective learners actively and meaningfully construct knowledge by assimilating new concepts under the guidance of teachers by building on their students' existing cognitive structures. Novak's research on concept maps in this context resulted in a very broad definition of concept mapping that can be reduced to mean "a structured way to organize information consisting of nodes ad their connections presented in a visually meaningful way." Given this general definition, advanced concept maps in STEM education, can range from strictly logical maps such as Venn diagrams to the highly complex and domain specific concept maps used in nursing education along with other examples covered in this book including diagrams and models that educators and others would not ordinarily think of as "concept maps."

Section 1 of the book consists of a concept maps history written as an empirical exercise to determine their foundations, for as Newton said, Google would reiterate, and constructivists would agree, all new knowledge of significance is built on the shoulders of giants. In Section 2, it was found, thanks to cognitive styles psychology, that concept maps can be classified into three types: object maps, verbal maps, and spatial maps corresponding to three distinct styles of learning and communication. According to neuropsychologists Olysa Blazhenkova and Maria Kozhevnikov, object learners and communicators are found among artists and multi-media persons who process information through very colorful, concrete, multi-dimensional, and multi-sensory images. The verbal style of communicating and processing of information, according to the media scholar, Marshall McLuhan, has dominated Western learning for centuries, especially after the invention of the Gutenberg press and the printed book in phonetic alphabet format. This cognitive style is opposed to the object style and a third type, spatial style in that spatial learners, as in the case of object learners, process information non-verbally, and through images.

Spatial thinkers tend to be less verbal and more visual and are found among scientists, engineers and mathematicians. The spatial cognitive style, however, differs from the object style in that spatial processing involve images that are highly abstract, geometrical, and logical. Neuropsychologists also claimed to have found that the different cognitive modes have their origins in different regions of the cerebral cortex with object cognition activating regions in the right brain, verbal learning locations in the left, and spatial cognition in a region in between the two hemispheres. Educators, who tend to prefer verbal learning, need listen more to what scientists, especially the psychologists, have to say about how we learn to gain a better understanding of using concept mapping in their domain.

Advanced Concept Maps in STEM Education: Emerging Research and Opportunities is essentially a Western book, written with a Western orientation. Had it been written in another culture, this alternate book might have told the story of concept mapping with a different emphasis such as more definitively hypothesizing that the future of concept mapping, especially advanced concept maps, belongs to spatial maps. This alternate book would also emphasize that the origins of spatial maps can be found in an art form known as origami and give greater detail on the history of using paper folding as a geometric mapping tool. Such a tool can be traced back to the ancient Japanese who painted *sangaku,* or mathematical problems on wooden tablets hung in shrines from 1600 to 1890. That *sangaku,* which included paper folding as a mathematical technique to organize and communicate information, were hung in Buddhist shrines also show that this alternate book, while differing in details, will have parallels to this book. Specifically concept mapping had religious roots in the East as well as in the West, as illustrated by Ramon Lull's maps in Majorca during the Middle Ages.

A parallel but similar version of this book would moreover relate that in 1893 an Indian mathematician, T. Sundara Rao published a book titled *Geometric Exercises in Paper Folding,* which related how to use origami, to demonstrate proofs of geometrical constructions, including an approximate trisection of angles that may have been misconstrued to imply that that the construction of a cube root was impossible. Sundara Rao's book and the idea of origami as mathematical models was transmitted to the west when the concept was mentioned by Felix Klein in one of his popular books on mathematics. Margharita P. Beloch, an Italian mathematician, perhaps influenced by the German, Klein, showed that the eponymous 'Beloch fold' could be used to solve Huzita–Hatori axioms, formulated by a Japanese mathematician and a Japanese origami artist. Solving the Huzila-Hatori axioms_resulted in the first application of origami to solve a general cubic equation.

The concluding chapter of *Advanced Concept Maps in STEM Education* raises the anomaly of object concept maps being orthogonal to spatial concept maps and suggests that the solution of this anomaly would be instrumental in a significant advancement of concept mapping in general. This preface suggests that origami geometric mapping may be the solution to the anomaly posed towards the end of this book and the future of concept mapping will be in non-verbal spatial 3D diagrams in the form of paper folding. Finally, this book is dedicated to my wife, Janice Tang, our two children, Elizabeth Polk, a physician and Matthew Tang, a geodetic computer engineer and to David Hyerle, educator in the best sense of the word, inventor of Thinking Maps and a person whose informal collaboration with me on concept maps helped make this book possible.

In addition, this book was partially made possible by funding from the National Science Foundation (NSF). Award Abstract #1505074: Sustainable STEM Learning Program (S2LP): Promoting Systems Thinking to Aid Holistic Education.

Section 1

A History of Concept Maps as Precursors of Modern Organizational Maps

Concept maps, a subspecies of cartography maps, are organizational diagrams that lay out the geography of the mind instead representing physical places. The origins of this organizational device can be found, among others, in the work of Raymond Lull in the middle ages, Peter Ramus during the Protestant Reformation and Leonhard Euler in the 18th century. Throughout this Insight Book the authors will refer to the term "concept maps," as maps associated with its inventors, Joseph D. Novak & Alberto J. Cañas and as a shorten term for "conceptual maps," which is a more general term than other such maps as Barzun's Mind Maps, Novak's concept maps and Hyerle's Thinking Maps®. All of these maps can be regarded as tools that educators, instructional designers, engineers, technical writers and others can use to organize, structure, store and present information and knowledge.

Chapter 1
A Brief History of Concept Maps

ABSTRACT

This section of concept maps will trace the works of Ramon Lull and Peter Ramus by comparing and contrasting their respective diagrams. In retrospect it will be argued that Ramon Lull's maps, which were basically religious in nature, were what we could call today, object maps, because of their colorful, concrete, almost multi-sensory presentation. On the hand, Peter Ramus's maps, according to cognitive styles theory to be discussed in a later chapter, were comparatively highly verbal and abstract. Marshall McLuhan, whose ideas lurk in the background throughout this book would say that according to his "law", "the medium is the message," Lull's medium, the illuminated manuscript on vellum, was quite different that Ramus' medium which was the printed book. The chapter ends with the philosophical language of Leibnitz, which has elements of what will be called verbal maps and spatial maps, two definite trends in concept mapping.

RAMON LULL

Historically, conceptual maps can be traced back at least to the Middle Ages, when such maps were used as memory devices for religious purposes and as educational tools to show others "to find their way" among spiritual to secular truths deemed important at that time. The predominantly religious aspect of the medieval maps can be seen in figure 1 which today could be

DOI: 10.4018/978-1-5225-2184-6.ch001

Figure 1. Spiritual mind maps in the Middle Ages and Renaissance. Source: http://2.bp.blogspot.com/-Ngpojitzqpg/VlegmgcLU0I/AAAAAAAAG-w/ipLRZW0Eeug/s1600/joachim-lull-q-branch.jpg

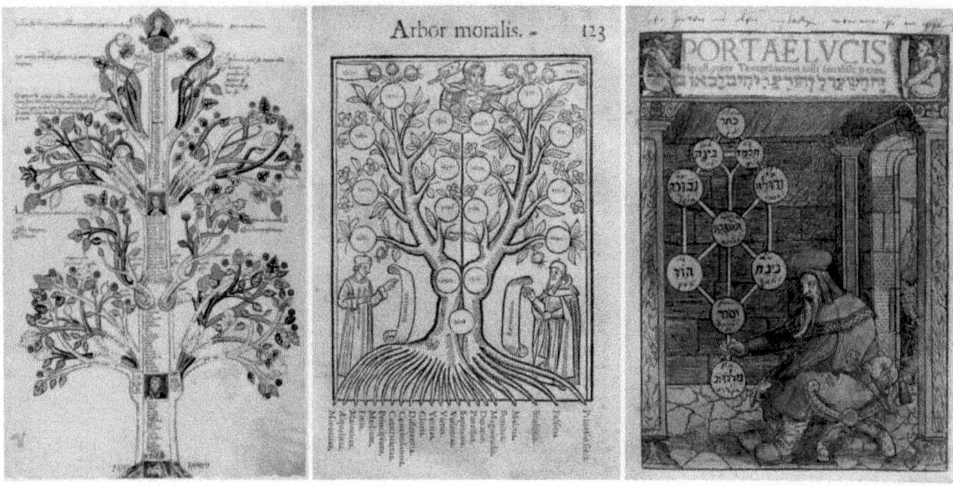

characterized as mind maps to help the faithful in an age when spiritual knowledge was more important than scientific other forms.

Perhaps the best known of these spiritual mind maps is found in the work of Ramon Lull, who designed a dynamic diagram which could be regarded as the first logic machine ever by taking popular trees of life illustrations and converting them to a religious machine, which he called, the *Ars Magnas.* According to Martin Gardner, mathematician and popular science writer an important history of science book, *Logic Machines and Diagrams,* Lull constructed his maps to answer spiritual questions based on the religious authorities of his time. Although Ramon Lull's main motive for inventing his device was to use it to convert the heathen, mainly Muslims and Jews, to Christianity, he probably got the idea for the invention from the zairja, a device Muslim astrologers used to make predictions along with generating answers and commentary on astrological matters (Gardner, 1958).

Ramon Lull (1232 – c. 1315) was born in Palma, the capital of the newly formed Kingdom of Majorca. As part of the effort to colonize the Mediterranean island, Lull's wealthy aristocratic parents came from Catalonia of present day Spain, as occupiers of the island which the Catalonians had conquered, enslaving all the Muslim population who had not been able to flee the island. Although married with two children, Ramon lived his youth as a licentious and wasteful troubadour. Lull, however, could not have been a complete wastrel

Figure 2. An Arabic zairja from which Lull may gave gotten his oracle machine, the Ars Magnus. Image source: https://c1.staticflickr.com/4/3103/3126431215_03aec8ab0f.jpg

in that he was tutor to the future king of Sicily, James II of Aragon and later became the administrative head of the royal household (Yates, 2003). Then came the conversion of the type so loved by the Church and her followers.

In a series of visions in 1263 Lull experienced a religious epiphany which he narrated in his autobiography *Vita coaetanea, A Daily Life* (Lluh, 1985).

Ramon, while still a young man and Seneschal to the King of Majorca, was very given to composing worthless songs and poems and to doing other licentious things. One night he was sitting beside his bed, about to compose

Figure 3. Ramon Lull's Spiritual Conceptual Maps, Tree of Life (left) and Ars Magas (Right). Source: http://images.eldiario.es/turing/Arbol-ciencia-combinacion-Ars-Magna_EDIIMA20141120_0495_13.jpg

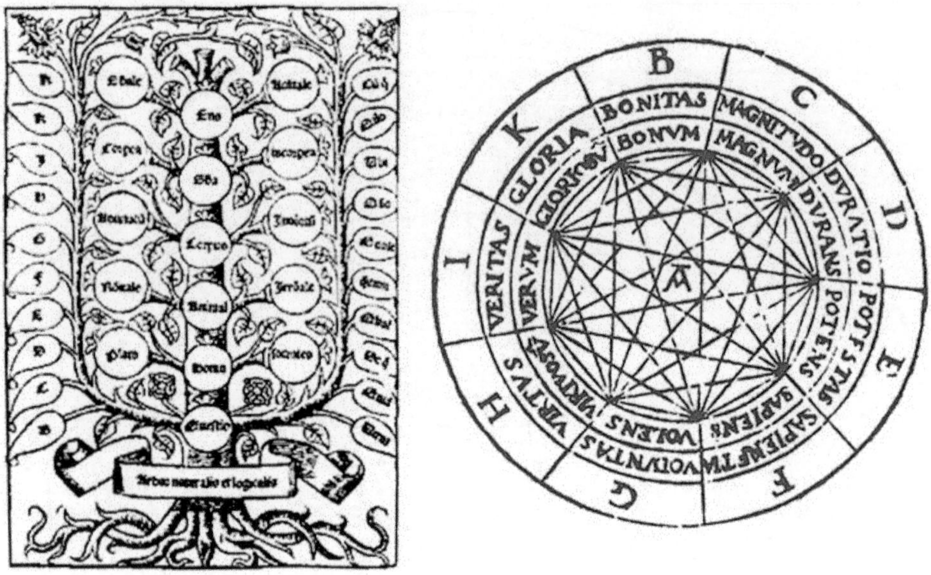

and write in his vulgar tongue a song to a lady whom he loved with a foolish love; and as he began to write this song, he looked to his right and saw our Lord Jesus Christ on the Cross, as if suspended in mid-air (pp. 10-11).

The vision came to him six times ending in Lull's forsaking family, position and possessions to pursue a life in the service of God, which included spending nine years of solitude during which time he worked to found religious institutions that would teach foreign languages and writing a book on how to overcome heathen arguments and objections to being converted to Catholicism. During the nine years of relative solitude and intense contemplation he read extensively in both Latin and Arabic including the works of the Muslim logician, Al-Ghazali's, who founded the science of optics. In 1274, while staying at a hermitage on Puig de Randa, the great book he intended to write, the *Ars Magnas,* originally entitled, *Art Abreujada d'Atrobar Veritat* or t*he Abbreviated Art of Finding Truth* was finally given to him through divine revelation, (Lluh, 1985).

The *Ars Magna*, "The Great Art," was a kind of highly structured Ouija Board through which an inquirer could get an answer to a question by turning the Great Art's concentric circles, which were broken up into concepts such as "truth", "glory" "goodness" (See Figure 3, second outer circle) which may be related to concepts in another circle such as, "the good," "the great", "the wise", "the powerful," (Figure 3: third inner circle). The outer most circle, consisting of the Letters B through K, representing the most important concepts when combined with the concepts of the other circles could answer questions ranging from the most spiritual to the most mundane. For example, a king might ask the question, "How can I gain the confidence of God in the coming battle?" and by spinning the concentric circles, the magic oracle might answer, "The wise and the great must also be good if he is to be truly powerful." According to Martin Gardner, quoting from Lull, the logic behind arriving at such a conclusion appears to have been highly complex.

If in Thy three properties there were no difference . . . the demonstration would give the D to the H of the A with the F and the G as it does with the E, and yet the K would not give significance to the H of any defect in the F or the G; but since diversity is shown in the demonstration that the D makes of the E and the F and the G with the / and the K, therefore the H has certain scientific knowledge of Thy holy and glorious Trinity (Quoted in Gardner, 1958, pg. 5).

If we were to classify Ramon Lull's conceptual maps today, we would call them mind maps to answer religious and other questions. Although not as "logical," the best example we have in the west is psychological diagram, which Carl Jung may have retrieved from the Middle Ages.

The diagram to the left in figure 3 is the psychologist, Carl Jung's first mandala, which he called "The Scheme of the World" (Slegelis, 1988).

To the right is an example of a Buddhist mandala, used primarily to "organize" religious truths According to Stanford educator, Barabara Tverskey (2011).

Depictive expressions of thought predate written language by thousands of years. They have evolved in communities through a kind of informal user testing that has refined them. Analyzing common visual communications reveals consistencies that illuminate how people think as well as guide design; the process can be brought into the laboratory and accelerated. Like language, visual communications abstract and schematize; unlike language, they use properties of the page (e.g., proximity and place: center, horizontal/ up–down, vertical/left–right) and the marks on it (e.g., dots, lines, arrows,

Figure 4. Psychological mind maps: Jungian (left) and Tibetan (right). Image sources: http://www.buddhanet.net/images/scmandala.jpg and https://fecdn-fractalenlighten.netdna-ssl.com/wp-content/uploads/2013/05/carl-jung-first-mandala.jpg

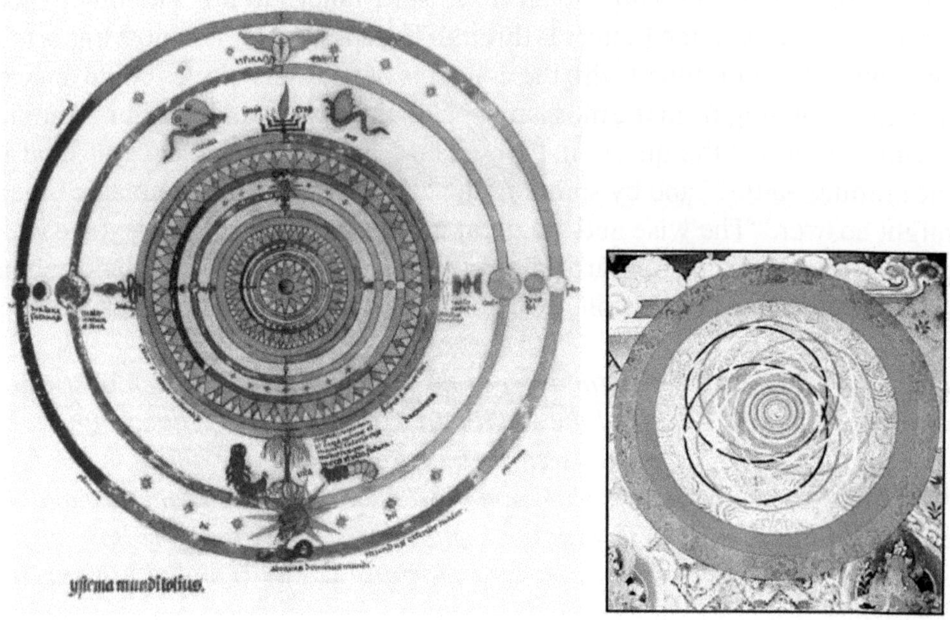

boxes, blobs, likenesses, symbols) to convey meanings. The visual expressions of these meanings (e.g., individual, category, order, relation, correspondence, continuum, hierarchy) have analogs in language, gesture, and especially in the patterns that are created when people design the world around them, arranging things into piles and rows and hierarchies and arrays, spatial-abstraction-action interconnections termed spractions. The designed world is a diagram (p 499).

Conceptual diagrams therefore predate written language, which in the West first appeared with the Phoenician alphabet, which, when adopted by the Greeks, according to Marshal McLuhan, completely changed the way Westerners thought. This change in thought as a consequence of the intrusion of new media, the phonetic alphabet, would culminate with the invention of the printing press, which conditioned us to think and perceive the world in a completely different manner. The best example of that change in our mental focusing lens, is found in the conceptual maps of Peter Ramus, who had a profound influence on education in Western Europe and later the United States in the 17th and 18th centuries.

As will be discuss in greater detail in later chapters, Lull's maps, according to cognitive style theory, could be regarded as object maps, which is a kind of map characterized by not so much by abstract logic but by emotional feeling. This kind of map, which in Ramon Lull's case was an attempt to organize spiritual, what we could call today, psychological feelings. While today we could consider such maps subjective, and therefore unscientific, this was not the way they saw things during Lull's time in the Middle Ages. For him and his fellow scholastics, being able to understand and grasp religious feeling, was a science, or theology, the study of God in all His manifestations.

Two centuries, later, however, theology, especially among the Protestants, expanded outwardly to emphasize the study of God's manifestations as exhibited in the natural world or in the European mind, which was becoming more and rational, and more concerned with the its logical operations. Peter Ramus, a key figure during the Sixteen century, was not only a religious reformer, but also an education reformer, who believe that the operations of the mind as they reflect the operations of later, could be "laid out" in in more rational manner --- that is, in a manner that would be more concerned with the intellect and not the emotions, as was the case with Raymond Lull. This change in emphasis from mapping out the soul to the mapping out of thought can clearly be seen by comparing Raymon Lull's maps with those of Peter Ramus.

PETRUS RAMUS

Approximately 200 years later, Petrus Ramus (1517-=1572), at the beginning of the Scientific Revolution, reintroduced the conceptual map to European Protestants during the period known as the Reformation. A study of the changing structure of conceptual maps throughout the middle ages, by comparing the two "maps" in figure 3, it can be seen that the diagrams have becoming more and more abstract, with less color, intricate design and concrete images such as people, roots, tree branches, leaves and flowers.

According to the media scholar, Marshall McLuhan (1994), after the Renaissance, with the invention of the Gutenberg press and printing, knowledge construction became more and more abstract and linear as if the mind were being structured by the printed book as template. In conceptual map constructing, rhetoricians and logicians such as Petrus Ramus got

Figure 5. An ornate medieval map versus a "modern" map printed from the Gutenberg press. Sources: https://s-media-cache-ak0.pinimg.com/236x/6e/28/2b/6e282b202bb853884595757ce59eaa06.jpg and Ong (2005)

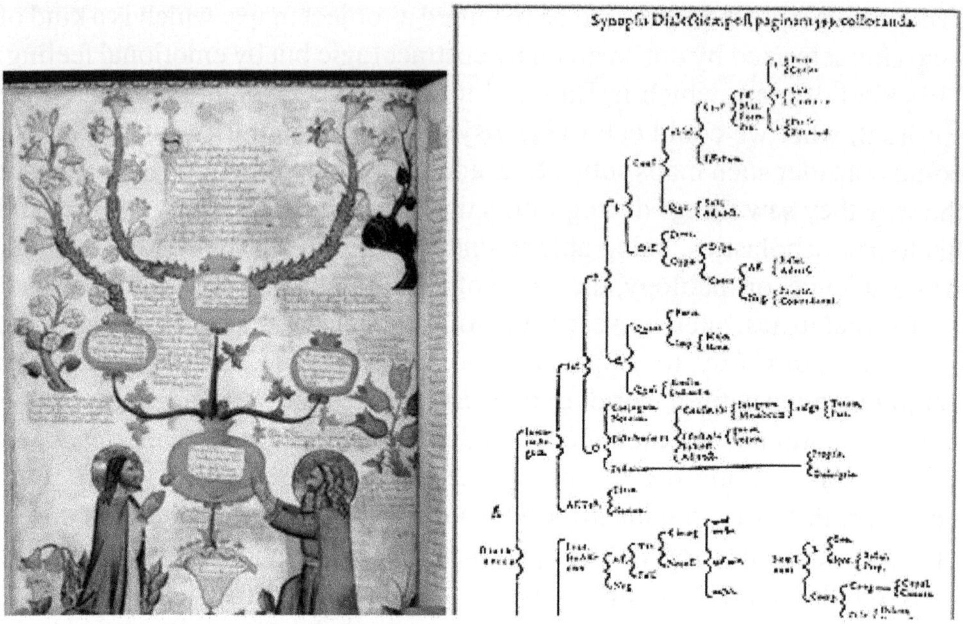

rid of ornamentation altogether and their maps consisted only of concepts themselves, often only as a single word, connected by lines, or brackets to indicate their relation connection to each other (Figure 3).

Petrus or Peter Ramus (1515-1572) was born at the village of Cuts in Picardy, France and was admitted to the Collège de Navarre, where he, being of farmer background, worked as a servant at the age of 12. At that time, in the 1520's, the universities were reacting strongly against Aristotelian scholasticism. According to his chief biographer, Walter Ong (2005), a Jesuit scholar, Ramus received his university degree in 1536, by writing a thesis entitled, *Quaecumque ab Aristotele dicta essent, commentitia esse*, which has been paraphrased as "All the things that Aristotle has said are inconsistent because they are poorly systematized and can be called to mind only by the use of arbitrary mnemonic devices.", which his followers simplified to "Everything Aristotle says is wrong" (36-41)

Ramus as a university graduate became a controversial teacher almost immediately upon getting his degree, engaging in public disputes with graduates and professors of the university. One opponent in 1543 was the Benedictine Joachim Périon, a renowned philologist and linguist who speculated on the origins of the French language (Schmitt, 2002). Ramus also publically argued with Jacques Charpentier, professor of medicine, who accused Ramus of undermining the foundations of philosophy and religion and another, Antonio de Gouveia, a lawyer, poet and classicist who flirted with Protestantism before returning to the Church. Because the debate with Gouveia was so acrimonious and of great public interest, their dispute was brought before the Parliament of Paris, and finally before the King, Francis I, who referred it to a commission of five. The commission sided with Gouveia and officially condemned Ramus, finding him guilty of arrogance and other charges and forbade him to continue lecturing. After his interdiction, Ramus withdrew from Paris, but after the decree against him was canceled by Henry II, who came to the throne in 1547, he returned to Paris and obtained a position at the Collège de Navarre (Mandrou, 1978).

In 1551 Henry II appointed Ramus as a professor at the Collège de France, where he taught philosophy and eloquence before audiences numbering as many as 2,000, in a time when a professor's income depended on the number of students he attracted. In spite of the professor's popularity his colleagues continued to attack him, among them a certain Pierre Galland, another professor who published a treatise entitled *Contra novam academiam Petri Rami oratio* ("Against the new academician, Peter Ramus, orator) in 1551, and called Petrus a "parricide" for his attitude toward Aristotle. The more serious charge in an age where authority was the final arbitrator of truth was the charge that Ramua was a *nova academia*, in other words, a sceptic and therefore to be denounced (Popkin, 1978).

Ten years later Ramus faced seven more enmity because of his adoption of Protestantism. He had to flee from Paris and, though he found an asylum in the palace of Fontainebleau, his house was pillaged and his library burned in his absence. He resumed his chair after this for a time, but in 1568 the sectarianism of Protestant versus Catholic was so intense and threatening that Ramus was advised to seek permission to travel, spending two years, in Germany and Switzerland (Ibid). When he returned to France, he got caught up in the St. Bartholomew's Day Massacre (1572), when Catholics were rioting against Protestant. Hiding for a while in a bookshop off the Rue St Jacques, he returned to his lodgings, on 26 August, the third day of the violence when he was captured and brutally murdered in his study at the Collège de Presles while at prayer and before being thrown into the Seine (Sellberg, 2016).

Influence

Ramus's anti-Aristotelianism arose out of a concern to reform and improve education as he and others believed that Aristotelian philosophy and education, in its current form as scholasticism was showing its age and was in a confused and disordered state. Ramus sought to infuse order and simplicity into philosophical and scholastic education by what he thought to be reinvigorating a sense of dialectic as the overriding logical and methodological basis for the various disciplines. When He published the *Aristotelicae Animadversiones* and the *Dialecticae Partitiones* in 1543, the *Animdadversiones* was intended to be a criticism on the old logic and the *Dialecticae* a new textbook of the science. It was in the "*Dialecticae partitiones*," that Ramus recommended the use of summaries, headings, citations and examples that took the form of Early Modern concept maps. Ong calls these maps as attempts to a reorganization of the whole of knowledge and indeed of the whole human lifeworld (Ong, 1994).

Although current scholars have described Ramus's conceptual logical diagrams, the attempt to "methodize" the arts, amateurish and simple-minded, Ramus enjoyed great celebrity with Ramists schools appearing everywhere with numerous adherents in France, Germany, Switzerland, and the Netherlands. Subsequently, The teachings of Ramus; teachings had a broadly based reception well into the seventeenth century and later movements, such as Baconianism, pansophism, and Cartesianism, in different ways built on Ramism, and took advantage of the space cleared by some of the simplifications (and oversimplifications it had effected. The longest-lasting strand of Ramism was in systematic Calvinist theology, where textbook treatments with a Ramist framework were still used into the eighteenth century, particularly in New England (Triche, 2004).

According to the well- known Jesuit scholar, Walter Ong, S.J., Petrus Ramus' concept maps were at the core of text book exegesis and was a major part of the Protestant religious, epistemological and education Reformation brought on and conditioned by the printing press. Walter Ong's work is of major interest to the history of concept mapping because of his theory on how different technologies affect the way we think, which is also a sub-thesis of this book. According to Ong, Ramus's conceptual maps represents the transition from the oral rhetorical illuminated manuscript culture of the Middle Ages to the modern text book age that had tremendous impact on theories of knowledge and how that knowledge is conveyed in education (Ong, 1994).

For many educators, scholars and philosophers in those countries most impacted by the printing press and the printed book, Ramus' method represented a new, more efficient way of organizing and communicating information in stark contrast to what these individuals, mostly Protestant, thought was the vague, over emotional, discursive, subjective and unscientific traditional methodologies of their adversaries both inside and outside the Catholic countries.

The Impact of Ramean Diagrams on Science

The "scientific" method of using Ramus' maps became so influential that in 17th Century, the natural philosopher and first president of the Royal Society, Bishop John Wilkins, tried to map out all known knowledge at that time in a book entitled, *An Essay towards a Real Character, and a Philosophical Language* in 1668 a year after Isaac Newton published his famous *Philosophiæ Naturalis Principia Mathematica,* which contained Newton's laws of motion and the universal law of gravitation, the book that formed he foundation of classical mechanics, containing laws that are still found in today's science and engineering textbooks.

The influence of Ramus and his method of diagramming knowledge visually in textbooks and in science cannot be over emphasized, For example, it is Ramus' mapping technique that laid the foundation of the biological sciences through Linnaeus invention of the binomial classification system and the foundation of chemistry through Dmitry Mendeleev (1834-1907) Periodic Table of the Elements. In the case of Carl Linneaus (1707-1798), the Swedish natural philosopher converted Ramus's tables with brackets and

tabulated the genus, species and families in a table that better fit the format of textbooks in his day as illustrated by figure 6. This rendering of maps to tables is another reason for the abandoning of concept maps as an organizing and presentation technique. As science became more and more popular and more and dependent on the textbook as the main instrument of study and learning, the printing technology made it easier publish information in a tabular format than as maps.

In the sixteenth century the idea of classifying all off nature in terms of classes and subclasses was a breakthrough in scientific thinking that was accompanied by an even greater scientific breakthrough, the use of diagrams as symbols that non-numerical knowledge could be calculated much like

Figure 6. Part of Linnaeus' classification system transposed from concept maps to fit the linear, more tabular format of an early biology textbook: Wikipedia. org_Wikicommons_images

mathematicians used mathematical symbols to formulate and discover knowledge of quantifiable things.

GOTTFRIED LEIBNIZ

Recent scholarship substantiated the influence of Ramus throughout the European continent if not across Atlantic to America. On the continent, one of the important personages Ramus influenced was Gottfried Leibniz, arch enemy of Isaac Newton, with whom he co-founded the calculus, whose notation has been widely used since its publication. Leibniz, however, was not at all impressed with Ramus' logic. What impressed the Gottfried, was the way his French predecessor had laid out information in tabular and diagrammatic form, a method that appealed especially to the commercial mind, and on a higher thought level to persons like Leibniz who wanted to make knowledge not only more accessible but more efficient, both in terms of its presentation and acquisition (Blank, 2008).

Gottfried Wilhelm (von) Leibniz born in 1646 and died in 1716 was truly a Renaissance man who lived after the Renaissance, polymath and philosopher

Figure 7. A specialized and modern Ramean concept map (right) of the evolution of bacteria, Archaea (spiders) and Eukaryota (plants) with a structure similar to Ramus' maps in the 16th century. Image source: http://creationwiki.org/pool/images/thumb/e/e7/PhylogeneticTree.jpg/300px-PhylogeneticTree.jpg

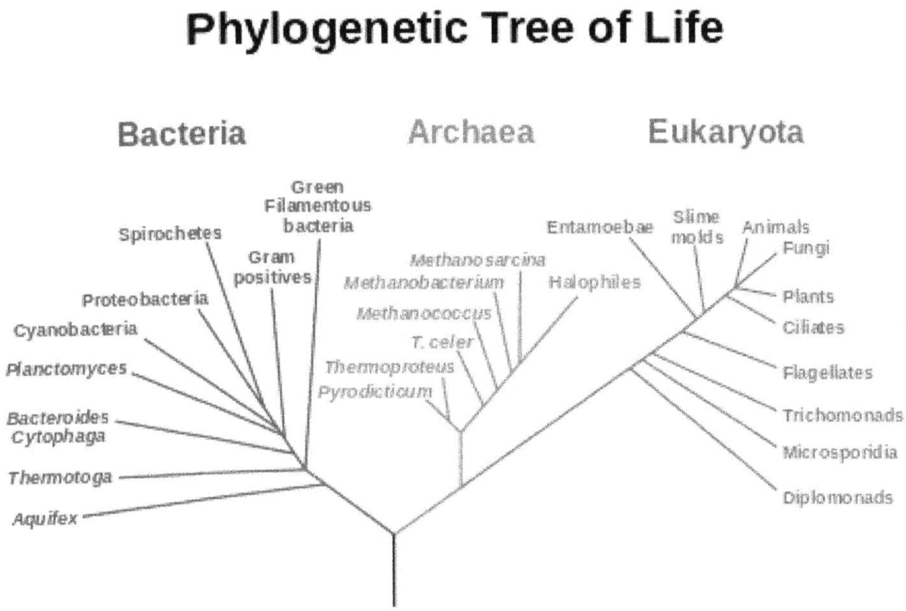

who holds a preeminent place in the history of logic, mathematics. In the sciences and engineering he was the first to describe a pinwheel calculator in 1685 and invented the Leibniz wheel, the first mass-produced mechanical calculator (Smith, 2011). He also refined the binary number system, which is the foundation of virtually all digital computers. Leibniz made major contributions to physics and technology, and anticipated notions that surfaced much later in philosophy, probability theory, biology, medicine, geology, psychology, linguistics, and computer science (Berlinsky, 2001). Leibniz's contributions to this vast array of scientific subjects, not to mention his contributions in philosophy were written in several languages, but manly in Latin, French and German were scattered in various learned journals, in tens of thousands of letters, and in unpublished manuscripts (Jolley, 1995).

In Leibniz's time, as it is today, it was difficult to separate logic from mathematics, and instead of believing that all problems could be best solved through reason, he believed that the best kind of reasoning was to calculate and was supposed to have said,

The only way to rectify our reasonings is to make them as tangible as those of the Mathematicians, so that we can find our error at a glance, and when there are disputes among persons, we can simply say: Let us calculate [calculemus], without further ado, to see who is right. (quoted in, Fabio, p 6)

Leibniz's calculus ratiocinator, which recent research has shown to early attempts at developing a symbolic logic systems, Leibniz's method to making such calculations outside the field of mathematics, which was to complement his other calculus, of mathematical equations (Couturat). For Leibniz, what Ramus had done was tabulated all know knowledge by reducing basic concepts to words, and ordered those words into an organized whole, with a logical structure, albeit as simple as that logic was. Moreover, because he thought symbols were essential to human understanding, he believed that the next step in making knowledge clearer and more "mathematical" by encoding words and concepts into a standard notation that would make calculation easier and more efficient. The result of this was the polyglot's call for a universal language and a universal real character, the latter being a universal symbol system such as mathematics, defining a character as any written sign, he then defined a "real" character as one that represents an idea directly and not through another medium as he thought was the case with European languages, which for based on the phonetic alphabet (Cohen, 1959).

Mathematical symbols are real characters or symbols because they point directly to the concept. Words as symbols in language based on the phonetic alphabet, such as Latin, English, French and German, were not real characters or signs because they pointed directly not to a concept but to a spoken word, which then pointed to a concept. For example, the spoken word "dog" points to a sound in English, a unit, which the linguists call a phoneme, not to the concept of a loyal, lovable creature, that one loves as a pet. On the other hand, the symbols, "1, 2. 3. 4 …, are real characters, which any literate person can read, whatever their first language, because they point not to sounds but to numerical concepts. Whether or not a written word points to a sound or a concept marks the difference between a phonetic language and an iconic language such as Chinese. It should be remarked that iconic characters as symbols are not confined to spoken languages and include such symbols as those found in astronomy, astrology, chemistry, Egyptian hieroglyphics, and in what we will call spatial concept maps (Courturat, 1969).

Around 400 years ago, Leibniz believed that if he could invent characters or signs suited for expressing all our thoughts as clearly and as exactly as arithmetic expresses numbers or geometry expresses lines, he could come up

Figure 8. Universal Affirmation modified from http://www.iep.utm.edu/leib-log/, where all B is C and all men are animals

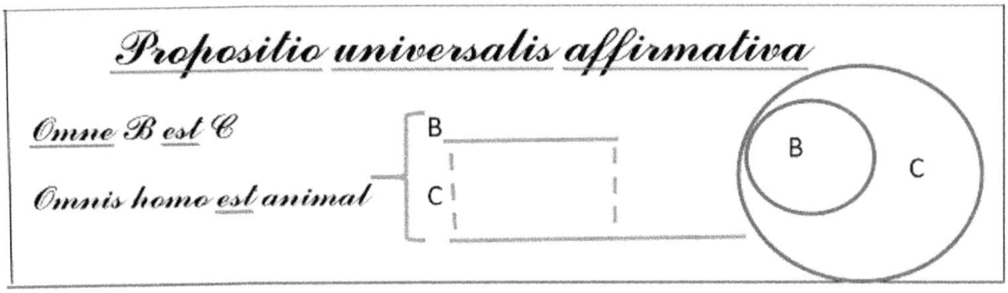

with a meta-cognitive machine that would be able to calculate thoughts, ideas and concepts as clearly, precisely and objectively as mathematics calculates its concepts in its domain. This new universal language would not only consist of a set of characters or standard iconic symbols, but also would also require something to hold the characters together to give the language meaning. That "something" was a syntax, which Leibniz found in symbolic logic, something he would have to invent just as he invented the mathematical logic behind the calculus. It was in the developing of this new syntax, which would later materialize in a formal logic system, that Leibniz used non-verbal and non-linear concept maps. The logical form that Leibniz was working on at that time was the syllogism, which was known in the Middle Ages, which few students or even scholars could grasp, but which Leibniz had already worked out in a non-verbal and linear fashion as follows Couturat (1961, . 292-295):

The UA (Universal Affirmation), 'Every man is an animal', "Omne B est C," can be represented as "Every A is B" or by two horizontal *lines or by two nested circles both* which symbolize that the extension of B is *contained* in the extension of C (IEP, 2016).

As with figure 8, figure 9, universal negation is represented by two "characters" or diagrams, a set of lines and a set of circles, the latter of course

Figure 9. Universal Negation. Source: modified from http://www.iep.utm.edu/leib-log/

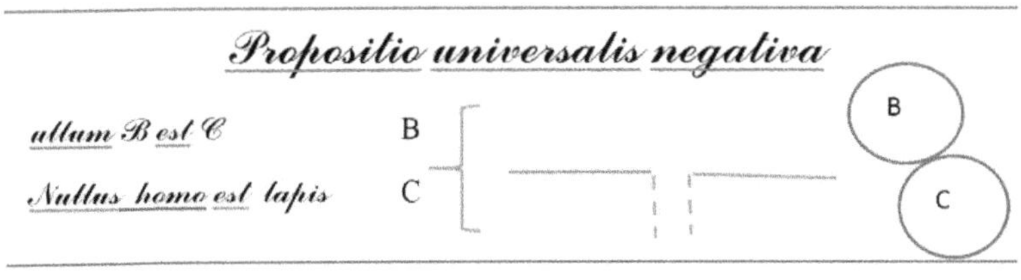

Figure 10. Particular affirmative propositions: "some B are C" and "Some men are wise", expressed as line and Venn diagrams and . Source: http://www.iep.utm.edu/leib-log/

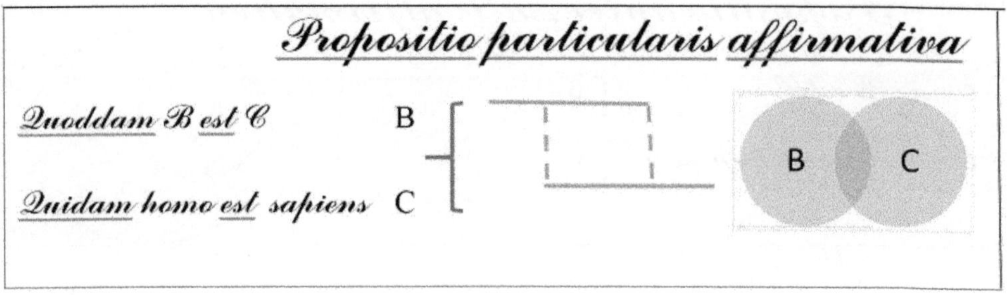

resembles and are precursors to Venn diagrams, which state that "no B is C" and "no men are rabbits."

Similarly, the following circles and lines symbolize that, in the case of a Universal Affirmation, such as 'some men are wise", the extensions of B and C overlap as per figure 10.

What is particularly remarkable about Leibniz was that not only was he highly advanced in his understanding of symbolic logic, he was able to express that logic in three different forms, verbally, and in two kind s of spatial conceptual maps, with line diagrams and in what are known as Euler and Venn diagrams

CONCLUSION

In this chapter we presented the verbal concepts maps of: 1) Raymond Lull, which were religious maps characterized by the medieval penchant for concrete, colorful, almost multi-sensory images with verbal labels. These medieval maps, hand drawn on expensive vellum and accessible to only a few such as the rich feudal aristocracy were what we might call today dynamic object maps that appealed to the emotions as well as the intellect; 2) Peter Ramus, which consisted mainly of bracketed concepts in words laid out in a diagrammed schema horizontally and hierarchically; and 3) Gottfried Liebniz's maps, which were remarkable in that he presented his logic through them in non-verbal spatial terms, both as line diagrams and circle diagrams which were basically Euler diagrams, which Leonard Euler would retrieve in also exactly the same form as Leibniz had constructed them a hundred years earlier. It might be noted that Raymond Lull's maps were available only to a few mainly because of their expense and because most people at that time were illiterate while Ramus's verbal maps were ubiquitous and almost universally understood because of the newly emergent highly literate commercial, Protestant middle class. This chapter ends with the remarkable work of Gottfried Leibniz whose work concept maps not only improved on Peter Ramus's ---- although unlike Ramus's diagrams, Leibniz's maps were not at all well-known ---, but consisted of a totally new kind of maps, non-verbal, non-linear, spatial maps, long before they were rediscovered after the Scientific Revolution.

Historically we have seen that Raymond Lull diagrams are what we will call "object" maps in which the verbal labels and the iconic images within the image are of equal importance. Information, in his case, spiritual or religious information is organized according to a gestalt where the over-all impact of the message or messages communicated are greater than the sum of its parts, which in modern systems talk we would call, its nodes and its connections. Ramean concept maps on the other hand, the parts, which are words, labels or nodes are also as important as the syntax or how the words are connected. However, the syntax is a relatively simple of genus-species relationships from the most general (the word/s at the left of the charts or maps) to the more specific. Basically, Ramus' maps convey the logic "y is a subset of x," and from this logic, he proceeds to use his maps to teach students the content of world.

Further research on Leibniz and his work on what was then called a philosophical language and universal character shows clearly that he understood the distinction between a meta-language and its parts, which are the universal characters. In the context of concept mapping, this means that concept maps themselves are parts of a meta-language and as such could be considered sentences in that language. These sentences, in turn can be broken down into smaller parts which are their characters or connecting symbols. Ramus' maps consist mainly of only one such character or symbol, the bracket which connects to nodes as words or concepts to mean that B is a part of or a subset of A to create hierarchical tree from the whole to its constituent parts or from larger sets to smaller sets. Later taxonomists such as Linneaus, the Swedish naturalist, and Demitri Mendeleef, the Russian chemist, would reorganize the parts of their sciences to tables, which better fit the format of the printed book.

REFERENCES

Baron, M. E. (1969). A note on the historical development of logic diagrams: Leibniz, Euler and Venn. *The Mathematical Gazette*, 113-125.

Buzan, T. (2006). *The ultimate book of mind maps: unlock your creativity, boost your memory, change your life*. HarperCollins, UK.

Charles, B., & Schmitt, C. B. (1992). Aristotle and the Renaissance. Cambridge, MA: Academic Press.

Cohen, J. (1954). On the project of a universal Character. *Mind, 63*(249), 49–63. doi:10.1093/mind/LXIII.249.49

Couturat, L. (Ed.). (1961). *Opuscules et fragments inédits de Leibniz. Presses universitaires de France. Reprint Hildesheim*. Olms.

Couturat, L. (1969). *La logique de Leibniz* (Vol. 38). Georg Olms Verlag. Blank.

Couturat, L. (2008). Ramus and Leibniz an analysis. InDaskal, M. (Ed.), *Leibniz: What kind of rationalist?* (pp. 155–166). Springer Netherlands.

Fabio, R. O. D. A. (2013). *Integrating high-level requirements in optimization problems: theory and applications* (Doctoral dissertation). Université Paris Nord, Paris, France.

Fei, L. J., & Sujan, D. (2013, January). Origami theory and its applications: A literature review. In *Proceedings of World Academy of Science, Engineering and Technology* (No. 73, p. 1131). World Academy of Science, Engineering and Technology (WASET).

Fish, A., Flower, J., & Howse, J. (2005). The semantics of augmented constraint diagrams. *Journal of Visual Languages and Computing, 16*(6), 541–573. doi:10.1016/j.jvlc.2005.03.001

Gardner, M. (1958). *Logic machines and diagrams*. McGraw-Hill.

Hull, T. C.Thomas C. Hull. (2011). Solving cubics with creases: The work of Beloch and Lill. *The American Mathematical Monthly, 118*(4), 307–315. doi:10.4169/amer.math.monthly.118.04.307

IEP. (2016). *Internet Encyclopedia of Philosophy*. Available at http://www.iep.utm.edu/leib-log/

Jolley, N. (Ed.). (1995). *The Cambridge companion to Leibniz*. Cambridge University Press.

Lemon, O. J. (1997). Logic and visual information. *Journal of Logic Language and Information*, *6*(2), 213–216. doi:10.1023/A:1008278914765

Llull, R., & Bonner, A. (1993). *Doctor illuminatus*. Princeton University Press.

LoLordo, A. (2006). *Pierre Gassendi and the birth of early modern philosophy*. Cambridge University Press. doi:10.1017/CBO9780511498503

Lovejoy, A. O. (2011). *The great chain of being: A study of the history of an idea*. Transaction Publishers.

McLuhan, M. (1994). *Understanding media: The extensions of man*. MIT press.

Ong, W. J. (2005). *Ramus, method, and the decay of dialogue: From the art of discourse to the art of reason*. University of Chicago Press.

Popkin, R. H. (1979). *The history of scepticism from Erasmus to Spinoza* (Vol. 226). Univ of California Press.

Rao, T. S. (1901). T. Sundara Row's geometric exercises in paper folding. Open Court Pub. Company.

Sellberg, E. (2016). Petrus Ramus. In *The Stanford encyclopedia of philosophy*. Available at: http://plato.stanford.edu/archives/sum2016/entries/ramus/

Slegelis, M. H. (1988). A study of Jungs mandala and its relationship to art psychotherapy. *The Arts in Psychotherapy*, *14*(4), 301–311. doi:10.1016/0197-4556(87)90018-9

Smith, J. E. (2011). *Divine machines: Leibniz and the sciences of life*. Princeton University Press. doi:10.1515/9781400838721

Tillyard, E. M. W. (2011). *The Elizabethan world picture*. Transaction Publishers.

Triche, S., & McKnight, D. (2004). The quest for method: The legacy of Peter Ramus. *History of Education*, *33*(1), 39–54. doi:10.1080/004676004 10001648751

Tversky, B. (2011). Visualizing thought. *Topics in Cognitive Science*, *3*(3), 499–535. doi:10.1111/j.1756-8765.2010.01113.x PMID:25164401

W. J. (2005). *Ramus, method, and the decay of dialogue: From the art of discourse to the art of reason*. University of Chicago Press.

Yates, F. (2003). *The occult philosophy in the Elizabethan age*. Routledge.

Yates, F. A. (2014). *Giordano Bruno & Hermetic tradition*. Routledge.

Chapter 2
The Scientific Revolution and Its Aftermath

ABSTRACT

Concept mapping during the latter years of the Reformation as represented by Peter Ramus, was basically a verbal affair. Although Ramus claimed to be a logician, his logical ideas, even in his time, were superficial at best. However, while Ramean logic may not have been very sophisticated, his concept diagrams were highly ambitious as philosophical and educational devices in that they claimed to be meta-cognitive, and general devices, to improve, what today would be called critical thinking. This chapter discusses how the ethos of the Scientific Revolution, the drive away from a qualitative universe to a quantitative one, driven by printing technology, changed the idea of meta-cognitive concept mapping to a very specialized area, the field of formal logic as evidenced by the work of Leonhard Euler, Lewis Carroll and John Venn.

INTRODUCTION

The Scientific Revolution was a pivotal period in Western history when a new paradigm, the Newtonian paradigm, the universal frame of reference that still rules today, replaced the old feudal, medieval Aristotelian paradigm. In the latter paradigm, the one in which Ramon Lull worked under, the universe was finite and a living organism where everything had its place as ruled by the natural law of the Great Chain of Being (Lovejoy, 2011). This dominant

DOI: 10.4018/978-1-5225-2184-6.ch002

concept during the Middle Ages, the *Scala Naturae,* the "stair-way or ladder of nature," is a concept originating with Plato, Aristotle, Plotinus, and Proclus and further developed during the Middle Ages, reaching its full early modern Neoplatonism during the Renaissance. The Chain of Being was a strict, religious hierarchical structure of all matter and life, believed to have been decreed by God. The chain starts from God and progresses downward to angels, demons, stars, moon, kings, princes, nobles, commoners, animals, trees, other plants, precious stones, precious metals, and other minerals (Tillyard, 2011). It was this Earth centered universe that Johanne Kepler, along with Copernicus (1473- 1543) and Galileo (1564-1642) sought to replace.

Conceptual maps of the type Ramon Lull promoted continued in Europe, but underground in the form of astrological maps, as the Church officially forbade the practice of astrology. These maps, however resurface during the Renaissance and may have been influential in shaping Johanne Kepler as one of the founders of the Scientific Revolution.

Johannes Kepler

Johannes Kepler (1571 –1630) was a German mathematician, astronomer, and astrologer and like the lesser known Petrus Ramus, a key figure in the 17th century scientific revolution. He is best known for his laws of planetary motion, still found in college texts on astronomy as these laws were pivotal for Isaac Newton's theory of universal gravitation. Kepler was a mathematics teacher at a seminary school in Graz, Austria, where he became an associate of Prince Hans Ulrich von Eggenberg and later an assistant to the astronomer Tycho Brahe, from whom he obtain the empirical data needed to prove his planetary laws. Eventually he was the imperial mathematician to Emperor Rudolf II and his two successors Matthias and Ferdinand II, a mathematics teacher in Linz, Austria, and an adviser to General Wallenstein. Additionally, he studied optics, invented an improved version of the refracting telescope (the Keplerian telescope), and was mentioned in the telescopic discoveries of his contemporary Galileo Galilei, with whom he corresponded regularly (Casper, 1959)

Kepler lived in an era when astronomy and science astrology existed side by side and both regarded as sciences with the common language of geometry, which was also the scientific language of astrology. Kepler also incorporated religious arguments and reasoning into his astrological and astronomical work, motivated by the religious conviction and belief that God had created the world according to a divine but rational plan that is accessible through the natural light of reason. Kepler described his new astronomy as celestial physics, an excursion into Aristotle's Metaphysics and as a supplement to his *On the Heavens* (Barker, 2001) transforming the ancient tradition of physical cosmology by treating astronomy and therefore astrology as part of a universal mathematical physics. (Korye, 1973).

That Kepler, one of the founders of the Scientific Revolution practiced astrology was first made clear, by the historian of science, Arthur Koestler (1959), in his chapter on the astronomer/astrologer in which he described details of horoscopes, Kepler cast on his family. On his father, he wrote:

Heinrich, my father, born 1547, 19 January . . . A man vicious, inflexible, quarrelsome and doomed to a bad end. Venus and Mars increased his malice. Jupiter combust in descension made him a pauper but gave him a rich wife. Saturn in VII made him study gunnery; many enemies, a quarrelsome marriage . . . a vain love of honors, and vain hopes about them (p 229).

Although both astrology and astronomy used the common language of geometry as understood in its day, what distinguished one discipline from the other is interesting. Astronomy, with its main tool, mathematical conceptual maps, which today we see as the objective study of the planets and stars, was then regarded as a spiritual and religious discipline, the "divine" science, while astrology was considered a mundane, vulgar practical tool, useful for purposes such as predicting the weather, determining the outcome of wars and the like. As a positivist scholar, Gerard Simon (1975), clearly stated:

Again, unlike us, who would be inclined to associate astrology with magico-religious thought, and astronomy with positivism, for Kepler it is astrology that is the profane utilitarian activity, while astronomy is the science of the sacred, the science of Creation. Neither his value judgements nor his motivation are like ours: which is an indication of the very great epistemological distance between him and us (p 447).

In actuality, Kepler, when a struggling student and teacher, disliked practicing astrology and would have much preferred to spend all his time practicing the pure science of astronomy, but cast horoscopes because that was his main source living (Koestler, 1959).

Regardless as to whether astrology is mundane or even a pseudo-science, it is of interest in this chapter, because the astrological chart is the best

Figure 1. Early modern conceptual map of the Copernican System which conflated astronomy with astrology. Source: https://s-media-cache-ak0.pinimg.com/564x/de/e4/7f/dee47f7c909e048d1afe85f2c342ebb7.jpg and http://2.bp.blogspot.com/-bKrxhEPCI5o/T8DyAuXt

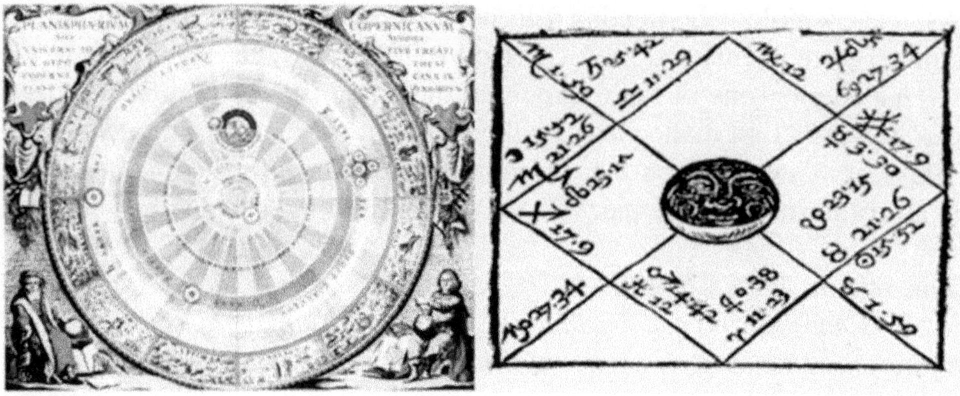

representative of a completely non-verbal concept map as can be seen in Figure 1.

Figure 1 shows two conceptual maps. The one to the left, an astronomical concept map of Copernicus' heliocentric universe, which resembles more of a spiritual or religious diagram, makes sense, because during the Renaissance and Scientific Revolution, astronomy was the "divine science," at a time when religion and science were conflated. On the other hand, the diagram to the right is a Renaissance astrological chart, which then was a secular tool used for mundane practical purposes, which much more resembles the more scientific maps that would follow the years after the Scientific Revolution. The astrological diagram in Figure 1, is more scientific in our eyes because it is much more abstract, consists of no images, except the ornamentation in the center and is completely non-verbal. The chart labels consist of only the glyphs for the planets, the sun and the moon: ☉, ☽, ☿, ♀, ♂, ♃, ♄, and the signs of the Zodiac, ♈, ♉, ♊, ♋, ♌, etc., which mark the zones of the diagram. According the standard textbook at that time, Ptolemy's *Tetrabiblos*, where and in what configuration the sun, moon and planets are located is extremely important as their location and how they relate to each other in space, in astrology, their aspects, determines the what has happened in the past, is occurring in the present and will occur in the past. It should also be noted the astrological chart, unlike the astronomical chart, not only is completely non-verbal, but the location of planets is accompanied by exact

Figure 2. the five Pythagorean solids. Source: http://whistleralley.com/polyhedra/platonic.gif

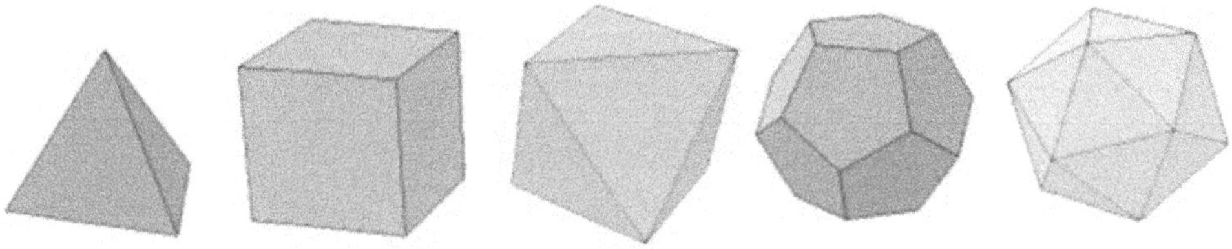

number as to position in the different zones. The astrological chart could be characterized as a highly visual, non-verbal, spatial, glyphic/iconic map as opposed to the highly verbal, hierarchical and linear Ramean maps, whose main purpose was pedagogical and a method of text exegesis or analysis.

Kepler's geometric images, which could be called very specialized concept maps, are also instructive not only to illustrate the their history, but can also reveal how concept maps work today. Kepler, unlike many of his contemporaries was an idealist in that he believed that the so called highest reality, was immaterial, and in his case, geometrical. Consequently he was attracted to the ideas of Pythagoras, who unlike his materialistic contemporaries who believed that the universe was made of some material substance such as air, earth, fire, water or atoms, asserted that the universe was made of number. Kepler interpreted that to mean that the universe was made up of geometric forms, and following Pythagoras, he maintained that these forms could be reduced to the five perfect immaterial solids as illustrated in Figure 2.

At this point, the reader should regard these five perfect solids as geometric symbols that will be used to organize information about thier natural world and the laws that operate in that world as did Kepler.

Consequently, the astronomer asked the question, "how can I create a map of the universe that will show that 1) the center of that universe is the sun and not the earth and 2) that universe operates under natural laws which are geometrical?" To answer, this question, Kepler nested the five immaterial perfect solids, or geometric forms, one within another in such a way that they if a mathematician drew a circle encompassing each of the perfect forms the result would be the map, or to use a better term, "conceptual model," found in Figure 1.

Figure 4, represented Kepler's final conceptual model as an imaginary mechanical construct.

Figure 3. Kepler's earlier conceptual model of the possible distances of planets from the sun as center of the system: https://images-na.ssl-images-amazon.com/images/I/51xhaE0VeML._SY344_BO1,204,203,200_.jpg

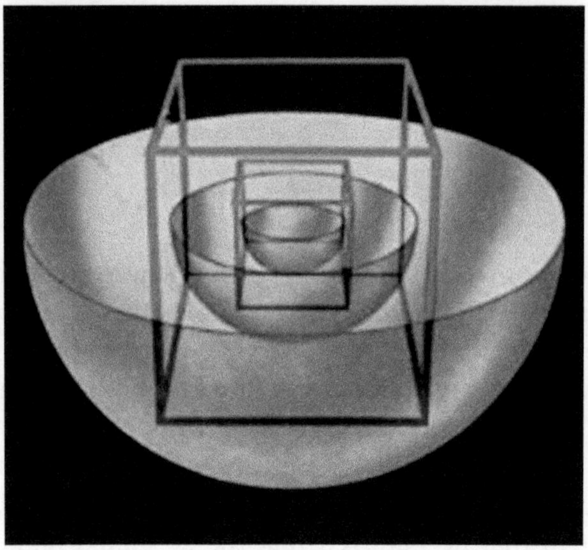

Figure 4. Geometric map of the sun centered universe with planet A, orbiting around the sun in an imaginary circle inscribed around the Pythagorean form, the square

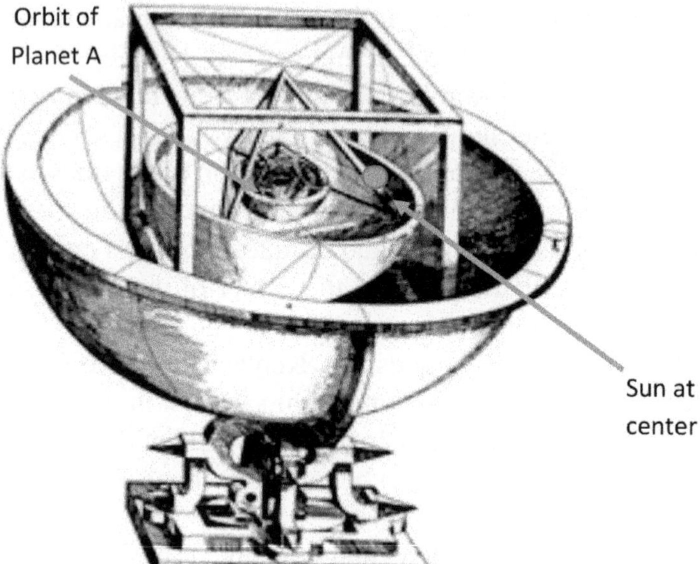

In the previous chapter we made a distinction between a universal language, which today we would call a meta-language and its constituent parts, which we identified in Ramus diagrams as made up of words and their connectors, the bracket. To continue with an analogy on Kepler's conceptual *models,* it can be said that the Pythagorean solids in figure 5, are the structural parts that make up Kepler's model of the universe, with its syntax, or the laws that hold it together are geometric ratios of the parts, the perfect solids, nested within each other. The distinction that we want to make here, is that all of the maps in the first chapter were concept maps of either thoughts (Ramus' maps) or emotions (Lull's maps).

Kepler's conceptual models are maps of the actual universe. Conceptual maps therefore consists of two kinds, 1) maps of the universe and the things in that universe, which are the conceptual models of the foregoing discussion of Kepler and 2) maps of thoughts and feelings, which were the maps discussed in the previous chapter. The diagrams discussed in the rest of this chapter, are "thought diagrams" or concept maps of the logical operations of the mind. These logic maps are similar to Ramus's diagrams in that they are conceptual models of thought processes, but are quite different in these maps are largely if not completely non-verbal. Moreover, these maps are radically different from Ramus's maps because the structure of the maps, including the relationship between the nodes, which can be labelled with but are usually represented as variables is preeminent in importance.

To summarize: Kepler's conceptual models are maps that point to the parts, structure and natural laws of the universe and its contents. In contrast, Ramus' diagrams are maps that point mainly to the content of scientific and other printed text while the logical diagrams of Leonard Euler (1707-1783), Lewis Carroll (1832-1898), John Venn (1834-1923) and Charles Sanders Peirce (1839-1914) are maps of the formal logical operations of the mind. These maps, unlike Ramus's point not to the content of concepts, but to their logical connections or syntax. Meanwhile, Ramon Lull's maps, which are diagrams related to spiritual matters, have been relegated to the dustbin of history by the Scientific Revolution, and will not reappear again, except in esoteric areas such as Jungian psychology.

Leonhard Euler

Leonhard Euler was born in Basel, Switzerland to Paul Euler, a pastor of the Reformed Church, and Marguerite née Brucker, a pastor's daughter. Soon after the birth of Leonhard, the Eulers moved from Basel to the town of Riehen, where Euler spent most of his childhood and where his family was close to the Bernoulli family, including Johann Bernoulli, Europe's foremost mathematician, father of Daniel Bernoulli, known for the Bernoulli Principle, a mathematical law that can be applied to and explain flight and mentor to the young Euler. Leonard he enrolled at the University of Basel. At that time Euler's main studies included theology, Greek, and Hebrew at he received a Master of Philosophy with a dissertation that compared the scientific ideas of Descartes and Newton (Finkel, 1897). During that time his father wanted Euler to follow him in becoming a pastor, but Johann Bernoulli convinced the father that Leonhard was destined to become a great mathematician. In 1726, Euler completed a dissertation on the propagation of sound but was unsuccessfully in obtaining a position at the University of Basel. Subsequently Leonard, left Basal for St Petersburg, Russia, where Bernoulli's two sons, including Daniel, were working. There Leonard found an academic position at the Academy of Saint Petersburg in the academy's math department. He lodged with Daniel Bernoulli with whom he often worked in close collaboration, both of them mastering Russian and settling into the intellectual life in Saint Petersburg (Debnath, 2009).

Euler Diagrams

According to Margaret Baron (1969) Leibniz produced diagrams before Euler similar to ones that follow, however, much of it was unpublished. She also observes even earlier Euler-like diagrams by Ramon Lull in the 13th Century as already discussed in Chapter 1, where we refer to Martin Gardner's (1958) interpretation of the history of concept maps as logical diagrams.

An Euler diagram is essentially a mathematical conceptual map that gives a visual representation of expressions in set theory. Take for example the math maps below in figure 5:

In the math map on the left, A, B and C represent three sets where A intersects B and C. Visually, Euler diagrams consist of *contours*, drawn as simple closed curves. The contours split the plane into *zones, which* can be identified by its containing contours. In the left diagram above, the contours (circles) are labelled A, B and C and the zones A, B, AB, AC and ABC are how they are related (Euler Diagrams, 2004). In the diagram to the right (figure 4), it can instantly be seen that while parrots are birds; and cats, dogs and cows, mammals, only, cats and dogs are usually considered as pets.

Figure 5. Two Euler Diagrams

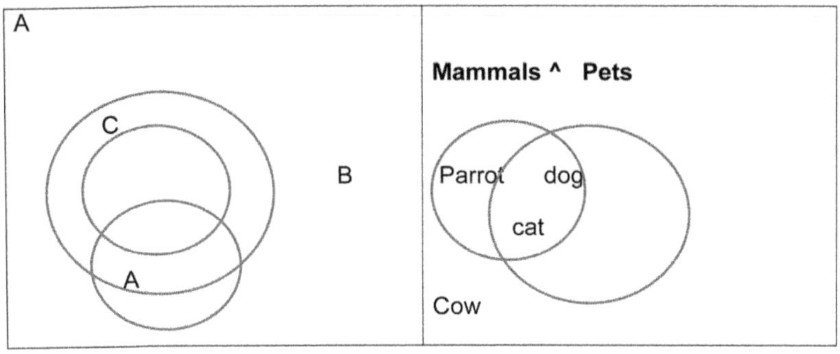

The original application of Euler diagrams as a way of diagrammatically demonstrating logic should be more widely used in schools where they are a great aid to teaching set theory and scholars are retrieving these diagrams as more than simply historical curiosities. For example, Eric Hammer has introduced a sound and complete logical system based on Euler diagrams attracting scholarly attention (Lemon, 1997; Stapleton, 2005). According to Shin (Euler Diagrams, 2004) who has developed a formal system.in this regard, shows that more expressive reasoning from Euler math maps can be achieved by extending the diagrams with graphs, which the Visual Modelling Group at the University of Brighton, along with others have done in producing are called spider (Howse, 2001) and constraint diagrams (Fish, 2005), as shown in Figure 6. These enhanced Euler diagrams can be seen as hypergraphs, or very advanced math conceptual diagrams and as such, show the power and potential of concept maps in general from the most colorful and concrete medieval spiritual maps to the highly abstract mathematical maps of Euler.

Figure 6 is an example of a modern day Euler unitary spider diagram that can be used in computer science and other domains, which in mathematics. The small black squares indicate the existence of an attribute described by the intersection of contours of two Euler diagrams. These points are joined together forming shapes that often look like a spider, connecting "or" conditions, also known as logical disjunctions. Thus figure 6 contains the logical statements, A or C or F and E, or G, or D, in which attribute C is linked to attribute F. The map also shows the following conjunctions: A and B, B and C, F and E and G and F (A ^ B, B ^ C, F ^ E, G ^ F). Moreover, in the universe of discourse defined by the concept map (i.e. diagram) in addition to the conjunctions specified above, all possible sets from A through B and D through G are available separately. The set C is only available as a subset

Figure 6. A modern day computer science Euler map. Image source: https://en.wikipedia.org/wiki/Spider_diagram

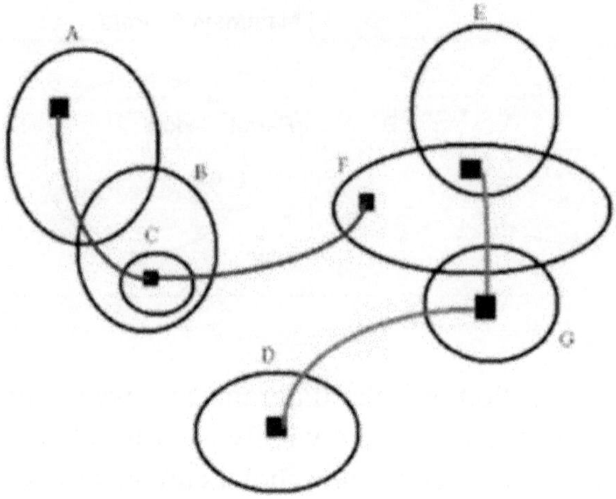

of B. The two "spiders" in the example include a blue spider indicated by the blue line and a red spider, the red line that connects D with G with E. When a temporal element is added to the diagram, an algorithm can easily be envisioned where after the red spider, (F ^ E) v (G) v (D) completes its operations, the result is then bed to the blue spider to perform its operations on the statement (A) v (C ^ B) v (F).

The history of contemporary spider diagrams can be traced back to Kepler's astrological charts and the idea of highly abstract non-verbal spatial charts with iconic glyphs sometimes but not always accompanied by numbers, had highly sophisticated origins during the Renaissance and the early stages of Scientific Revolution. Here, however, the difference between Kepler's conceptual models and Euler's should again be emphasized. Kepler's diagrams are maps of what is perceived to be the actual external reality, maps of the solar systems. Euler's diagrams are maps of logical operation which exist in the mind and may or may not have correspondence with actual things and events that take place in the real world.

John Venn

John Venn, who like Leonard Euler wanted to conceptualize formal logic with visual diagrams, was born in Kingston upon Hull, Yorkshire and like Leonhard Euler, of a religious parents. His father, the Reverend Henry Venn, was the rector of an English parish and descended from a long line of church evangelicals. The younger Venn began his education in London enrolling with his brother Henry in Sir Roger Cholmeley's School, now known as Highgate School, in September 1846. Later he moved to a proprietary school and in October 1853 went to Gonville and Caius College, Cambridge where obtained his tripos degree in mathematics, graduating sixth in the class, the highest student of his college. On the strength of this result he was a fellow of the college, the President of the College, a post he held until his death. In 1862 he became a lecturer in moral science, studying, teaching logic and probability theory, and giving intercollegiate lectures. (Edwards, 2004) These duties led to his developing the diagram which would eventually bear his name.

I began at once somewhat more steady work on the subjects and books which I should have to lecture on. I now first hit upon the diagrammatical device of representing propositions by inclusive and exclusive circles. Of course the device was not new then, but it was so obviously representative of the way in which any one, who approached the subject from the mathematical side, would attempt to visualise propositions, that it was forced upon me almost at once. (quoted in Edwards, pg. 3).

Venn diagrams, which are also called set diagrams or logic diagrams, are conceptual maps that show all possible logical relations between a finite collection of different sets. These diagrams depict set members points on a plane, and sets as regions inside closed curves. A Venn diagram consists of multiple overlapping closed curves, usually circles, each representing a set with the points inside a curve labelled *A* to represent the member of the set *A*, while points outside the boundary represent members, also called elements, not in the set A. For example, the set of all elements that are members of both sets A and B, A ∩ B, is represented visually by the area of overlap of the regions A and B.

In Venn diagrams the curves are overlapped in every possible way, showing all possible relations between the sets. They are thus a special case of Euler diagrams, which do not necessarily show all relations. Venn diagrams were conceived around 1880 by John Venn. They are used to teach elementary set theory, as well as illustrate simple set relationships in logic, computer science, probability, statistics, and linguistics.

Because they are so similar, Euler diagrams and Venn diagrams are often confused, and for good reason as their differences are highly technical ones. Venn diagrams can be seen as a special case of Euler diagrams --- i.e. Venn

Figure 7. A Venn diagram, consisting of A, B and C, with their intersections. http://www.gmatfree.com/wp-content/uploads/2014/07/ScreenHunter_159-Oct.-13-14.54.jpg

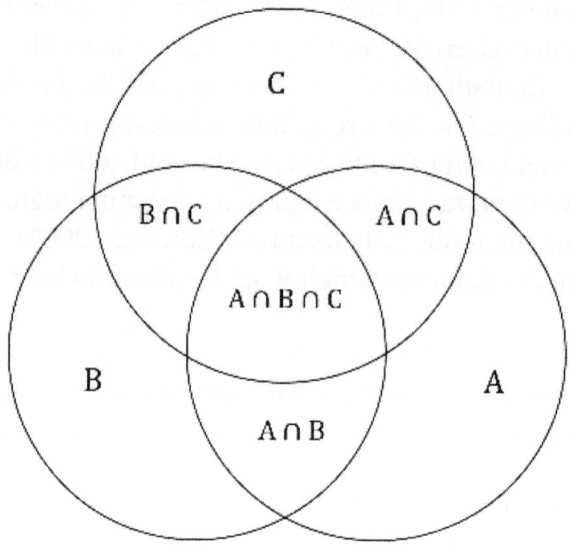

Diagrams are a subset of Euler Diagrams ---- as Venn diagrams (Venn: z = {a, b, c,}) must contain all possible zones, whereas Euler diagrams can contain a subset of all possible zones (Euler" x = {z}) Also, in Venn diagrams a shaded zone represents an empty set, whereas in an Euler diagram the corresponding zone could be missing from the diagram (Figure 5). This means that as the number of contours increase, Euler diagrams are typically less visually complex than the equivalent Venn diagram, particularly if the number of non-empty intersections is small (Euler Diagrams, 2004)).

Venn and Euler diagrams are so similar that John Venn, himself, called his own logical diagrams, "Eulerian Circles". Whatever their distinction and these non-verbal concept maps, were a diagrammatic means of representing sets and their relationships in which overlapping circles are used, in which the area of the shape is proportional to the number of elements it contains and were as conceptual maps were a step above the hierarchical tree map introduced by the Rameans. Figure 6 shows two conceptual maps, one Eulerian and the other Ramean, one using non-hierarchical circles and frames and other, an hierarchical tree, with both expressing the same statement that "animals and minerals or kinds of things and two kinds are animals are 4 legged ones and 2 legged ones." The Eulerian or Venn map, has a greater scientific advantage over the Ramean tree map, however, because the Venn diagram can easily be translated into a logical statement which can be formally proved or disproved as follows: Let T = Things, A = Animal, M = Mineral and 2 = 2 legged and 4 = 4 legged, then: A = {2 ^ 4}; M = {0}; 2 ^ 4 ⊄ M, etc. etc.

Whatever the slight differences between Euler and Venn logical conceptual maps, both had problems handling sets above three when the sets intersect, a problem which, apparently A. W. F. Edwards (Edwards, 1989) have solved. Edwards, a British statistician and evolutionary biologist devised a number of constructions for Venn Diagrams for five, six and seven sets which are widely recognized as easier to use than Venn's constructions for these dimensionalities. These, however, still present challenges to comprehension, partly as a result of using different shapes for different sets and the highly abstract mathematical nature of the diagrams. Nevertheless, the nested Venn Diagram has proven to be the nested Venn diagram, another name for Edwards' diagrams, is a useful alternative for visualizing and understanding the relationships between a number of sets, particularly when that number is between five and eight (Edwards. 2004)). As can be inferred from the substance of this chapter, the purpose of graphical representation is to aid speed of understanding, quality of understanding or both and the best diagrams allow the detection patterns that are not obvious from studying the numbers alone. Edwards makes a

Figure 8. A multi-set Venn diagram. https://quentinsf.com/static/media/uploads/Edwards-Venn-five.png

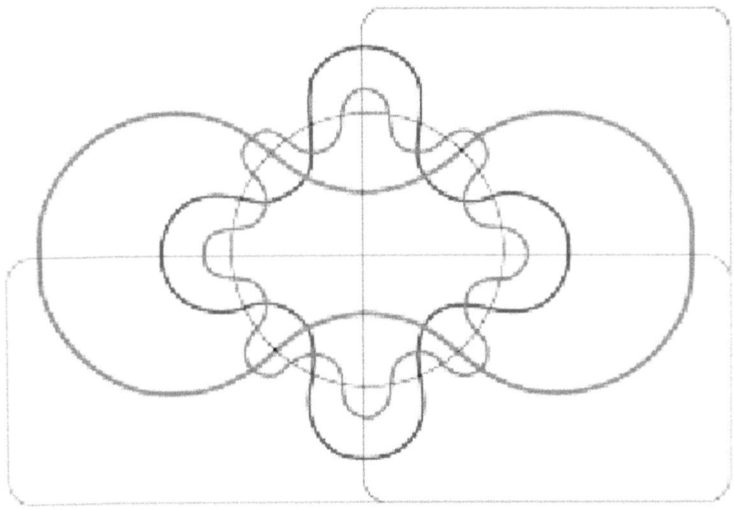

similar point, when he says (on page 47) "In mathematics a good diagram is an invaluable aid to clear reasoning whereas a bad one can seriously mislead."

Lewis Carroll

Lewis Carroll is much better known than Charles L. Dodson, which was his real name, also better known than John Venn because of Carroll's best known work *Alice in Wonderland.* Lewis or Dodgson, was a polymath who not only wrote the popular Alice books, but was also a mathematician and logician in his own right. Charles Lutwidge Dodgson, or Lewis Carroll, was born in the village of Daresbury, England, in 1832, the oldest son in a family of 11 children, with a cleric father, who with his wife, raised the children in a rectory. Carroll was rather adept at entertaining himself and his siblings, with all kinds of game including math games, where he excelled. During his early youth, Dodgson was educated at home. His reading lists preserved in the family archives testify to a precocious intellect: at the age of seven, he was reading books such as *The Pilgrim's Progress.* At the age of twelve, he was sent to Richmond Grammar School (now part of Richmond School) at nearby Richmond (Cohen, 1996).

In 1846, Dodgson entered Rugby School where he excelled as a student in mathematics. Dodgson, left Rugby at the end of 1849 and matriculated the next year at Christ College, Oxford, his father's old college. Again, as was the case at Rugby, although he was lackadaisical in his studies but academic achievement came easily to him earning first-class honors and Final Honors in the School of Mathematics and graduating first from college. He remained at Christ Church studying and teaching, but the next year he failed an important scholarship through his self-confessed inability to apply himself to study. Even so, his talent as a mathematician won him the Christ Church Mathematical Lectureship in 1855 which he continued to hold for the next 26 years until his death.

It was during his lectureship period at Christ Church that Dodson invented his Carroll Diagrams, named after his pen-name. These diagrams are basically visual and non-linear ways to learn to or represent the formal logic of sets, which is basically linear, when expressed in equation form using set theory symbols. For example, the following symbolic statements: 1) $A = \{3,7,9,14\}$ and $B = \{9,14,28\}$, means set A contains the numbers 3, 7, etc. while the Set B, 9, 14, etc.; or 2) $A \cap B = \{9,14\}$ means that set A intersects ($\cap$) Set b, or that the objects that belong to set A also belong to set B; while 3) the

Figure 9. An example of Dodson's photographic work: http://hrc.contentdm.oclc.org/cdm/landingpage/collection/p15878coll18#nav_top

statement, A ∪ B = {3,7,9,14,28} indicates the union (∪) of sets A and B, meaning that A and B. combined contain all the objects that belong to set A and set B. If you found this difficult to follow, so did Dodgson's students at Christ College (Wilson, 2008).

Usually, then as is the case now, student learn mathematics (and logic) by memorizing the statements and doing problems over and over again until they finally reach an understanding of what they have memorized through repetition. Professor Dodgson, thought there might be an easier, or at least less boring way to teach set theory, and did so by inventing an interactive logic game which consists of a conceptual diagram and tokens to move around in different places of the diagram, where different places meant different things (Wilson, 2008). In Figure 10 we have a Carroll 2 set diagram where the placement of the token indicates different set theory meanings, where A equals, say girls, and B equals taller than 5 feet.

In the above diagram, if you place the token at position W, you have the statement: W = A ∪ B, or W = girls who are taller than 5 feet, at another position X = A ∪ ~ B or, X = boys (i.e. not girls) who are taller than 5 feet; while Y = ~ A ∩ ~ B or Y is the set that includes only boys (~B)

The same statements can perhaps be easier said with a Venn diagram as in Figure 4, although what the minimalist Venn diagram gains over the Carroll diagram in expression, the Carroll Diagram may be a better map to teach set theory. Evidently the renowned mathematician, Stephen Wolfram,

Figure 10. A Carroll diagram to solve a 2 set logical problem

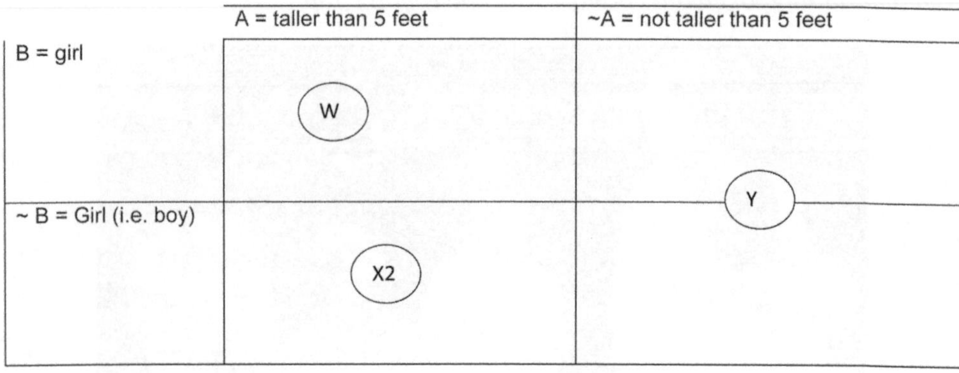

was of this opinion when he used the Carroll diagram instead of the Veen to as a teaching device for one of his Mathematica demonstration projects (Wolfram, 2016).

Lewis Carroll was an English writer, mathematician, logician, Anglican deacon, and pioneer photographer. His most famous fiction writing after *Alice's Adventures in Wonderland,* is its sequel *Through the Looking-Glass*, which includes the poem "Jabberwocky", and the poem "The Hunting of the Snark," all examples of the genre of literary nonsense that showed his facility at word play, logic and fantasy (Wilson, 2008). Dodson/Carroll cannot be considered a major figure in the history of logic diagrams as concept maps, although his work in that area remains formidable. His importance in this history, however, like Ramus, is that he was among the first to use visual diagrams to teach formal subject matters. This educational function behind concept mapping would again prove important in the 20th century. We now turn this chapter's attention to Charles Sanders Pierce, who is of great importance to the idea of concept mapping, in that he not only contributed to this area per se, he also provided one of the few theories on concept mapping from the perspective of semiotics, the study of signs and symbols and their use or interpretation.

Charles Sanders Peirce

Charles Sanders Peirce was born at in Cambridge, Massachusetts, the son of Sarah Hunt Mills and Benjamin Pierce, who was a professor of astronomy and mathematics at Harvard University and perhaps the first serious research mathematician in America. At age 12, Charles read his older brother's copy of Richard Whately's, *Elements of Logic*, a leading English-language text on the subject at that time. So began his lifelong fascination with logic and reasoning (Fisch, 1986). Between 1859 and 1891, Peirce was intermittently employed in various scientific capacities by the United States Coast Survey and its successor, the United States Coast and Geodetic Survey, where he enjoyed his highly influential father's protection until the latter's death in 1880. After his father's death, Peirce became destitute and The person one who did the most to help him in these desperate times was his old friend and colleague, William James, who dedicated g his *Will to Believe* (1897) to his mentor and arranging for his friend to give a series of paid lectures at or near Harvard from 1898 to 1910, the year of his death (Fisch, 1986, Burch, 2014).

Peirce was the American contemporary of the Brit, John Venn, but whereas Venn primarily practiced concept mapping, Pierce was also a theoretician who gave a better understanding of the philosophy or nature of the scientific or epistemological phenomenon of concept mapping. Specifically, Peirce firmly established the science of semiotics, the study of signs and symbols as communication processes by establishing a testable model of the relationship between the interpretant, who interprets a sign or symbol, which is somehow related to an object or event, . In figure a person sees smoke, interprets it as a sign of fire, and takes appropriate action, perhaps to check out whether or there is in fact a fire as indicated by the smoke, (Figure 11).

To take the same situation to higher level, to that of symbols and verbal thought, a reader (interpetant) would a read passage with the symbol string, "Where there's smoke, there's fire," (sign vehicle) in the context, of a rumor of, say, a politician, of wrong doing, and then proceed to read on to see if

Figure 11. Peirce's diagram of natural signs. Image source: https://decodingsemiotics.files.wordpress.com/2015/06/imgres.jpg?w=459&h=189

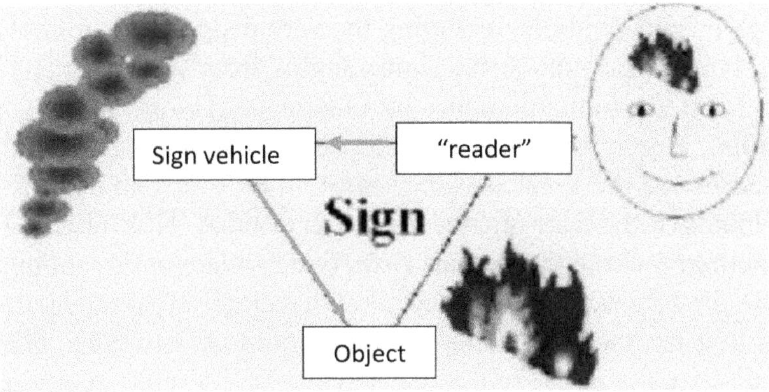

there is any truth to the rumor (object or event). In the context of concept maps, an interpretant would "read" the map and judge its value as to how well it represents the object or event that the map is supposed to represent (Kankunen, 2001, 2004).

C.K. Ogden the famous semanticist, and his colleague, I.A. Richards, came up with the same triatic semiotic model approximately 20 years later, in a famous book entitled, the *Meaning of Meaning* (Malinowski, 1923), which was called the "Triangle of Reference," which is a conceptual map which describes he triangle describes the relationship between the interpretant as subject, a concept as object or referent, and its designation (sign, signans.) In the example below (Figure 12), the subject sees the symbol, a no parking icon, and thinks, "don't park here" associating the sign as symbol its meaning. The same process is in operation when an interpret sees a concept map as symbol which has a reference, the object or event it represents, only on a much higher thought level than seeing smoke as a sign of fire and a parking sign as a warning not to park in a certain space. In the context of this chapter, the semiotics of concept mapping of course means working at a higher level of abstraction as a higher cognitive thought process, where a concept map (lower left of triangle) is a symbol, that symbolizes a thought or reference (apex of triangle) and also refers to or stands for a referent (bottom right), which can be a thought or an actual thing, such a chair, a car, or a space, where it is illegal to park.

Charles Peirce was also is sometimes known as "the father of pragmatism," having greatly influence fellow psychologist William James (1975). The scientist was educated as a chemist, but unlike many of his philosophical contemporaries, was not part of the higher education established and worked, mainly for the government, as a scientist for many years. Aside from being one of the founders of pragmatism, he could also be considered a founder of a field called semiotics, closely related to linguistics, which studies the structure and meaning of language more specifically. The semiosis is the study of signs and symbols, including the verbal signs and symbols used in language, but also the non-verbal signs ranging from smoke signals to music as a sound systems to mathematics as a language (Hoopes, 2014).

Although largely neglected until recently Bertrand Russell (1959) considered Peirce the greatest American thinker ever and one of the most original minds of the later nineteenth century and A. N. Whitehead, (1902) who co-authored wrote the famous *Principia Mathematica* said that he was struck how the American anticipated his own metaphysical and mathematical thinking. In a similar vein as these scientific and philosophical giants, Karl

Figure 12. Ogden's triangle of meaning, where a symbol symbolizes a concept which becomes a thought or reference and also stands for a referent. Image source:

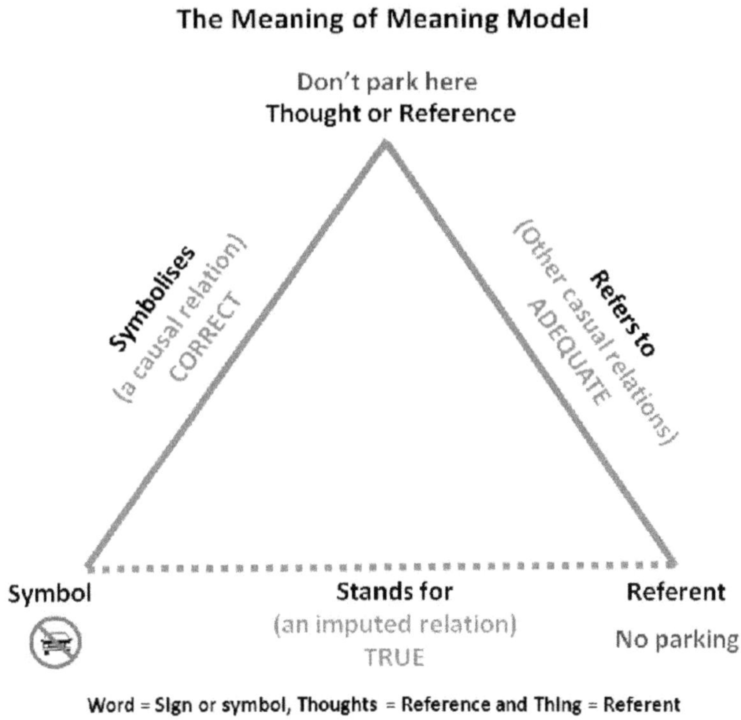

Popper, the philosopher of science viewed Peirce as one of the greatest philosophers of all times (Angelis, 1995; Moore, 1964).

One of the topics Peirce worked on throughout his life related to concept mapping was his existential graphs, which he began in early in his career and which he classified into three types, Peirce proposed three systems of existential graphs: *alpha* graphs which are isomorphic to sentential logic *beta*, which are isomorphic to the predicate calculus and *gamma* graphs which were nearly isomorphic to normal modal logic . For alpha graphs, the syntax includes the blank page, which Venn called a blank frame; single letters or phrases written anywhere on the page as labels.

Any graph constructed with this syntax be enclosed by a simple closed curve called a cut or sep. A cut can be empty and can nest and link at will, but must never intersect. For these graphs the semantics are The semantics are the blank page or frame denotes true, while an empty cut denotes false, as in boundary math where () = *, = False and if (()), then () = false (Bricken, 2016). Letters, phrases, subgraphs, and entire graphs may be True or False

Figure 13. Example of alpha graph: https://en.wikipedia.org/wiki/Existential_graph

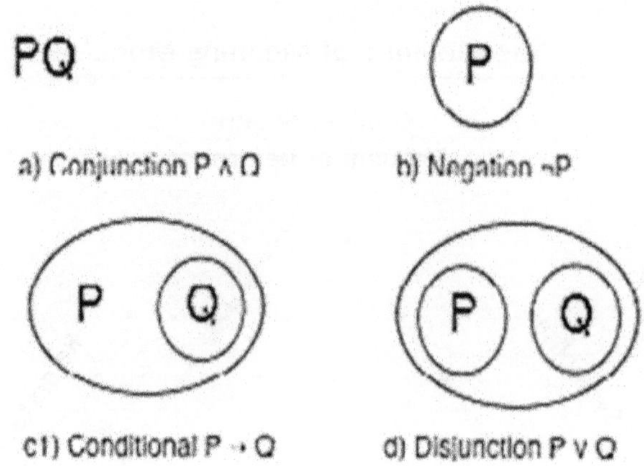

and and to enclose a subgraph with a cut is equivalent to logical negation or Boolean complementation. Accordingly alpha graphs are a visual minimalist notation for sentential logic, grounded in the expressive adequacy of the And (^) and the Not (¬) (Hammer, 1998). In structure and presentation these logical/mathematical spatial maps should be clearly separate and distinct from Ramean, verbal, linear and hierarchical maps, including text in traditional formats (figure n, right) as illustrated in the figure below. Ordinarily such text formats are not considered maps unless we expand its meaning to mean any collection of symbols, in this case, written words, organized spatially across an empty space, which in a this case is the empty page (Brown, 1969). The figures that follow show how the structure of textual concept maps have changed since Ramus' time, when the educator used such maps to teach the content of his day.

During the first stage of conceptual mapping since the Renaissance, Ramus and his followers, mapped out concepts in the form of bracketed diagrams (Figure 14).

In the second stage of textual diagrams, the text is rearranged as tables as illustrated in Figure 15's reproduction from a page in Linnaeus' taxonomy of animas.

Two centuries after Ramus introduced his diagrams to help students remember and understand the content of text in books, the diagrams disappear altogether (Figure 16).

Figure 14. First stage, Ramus' graphs. Image source: https://rcfrey.files.wordpress.com/2012/09/ramus1.gif

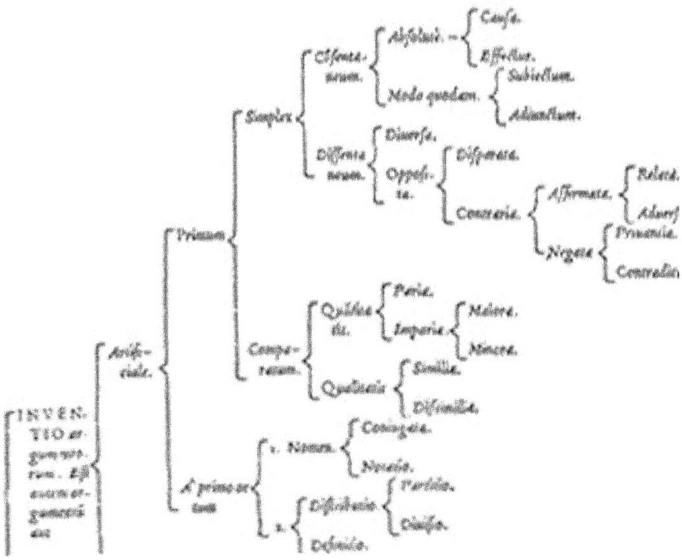

Figure 15. Second stage concept maps as tables such as Linnaeus' taxonomy of animals. Image source: https://en.wikipedia.org/wiki/Linnaean_taxonomy

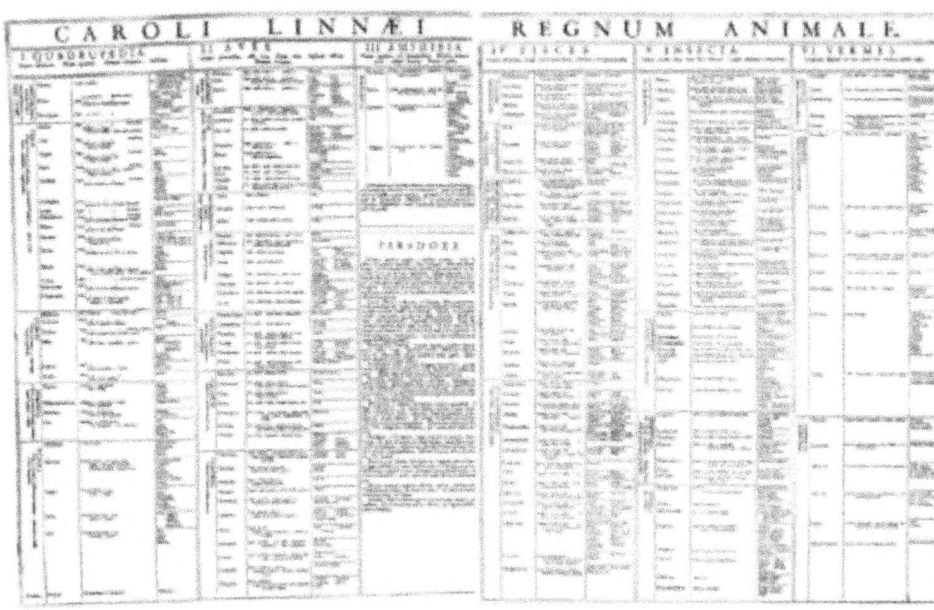

Figure 16. The third stage of Ramean maps in which diagrams disappear altogether replaced by pages in a printed book

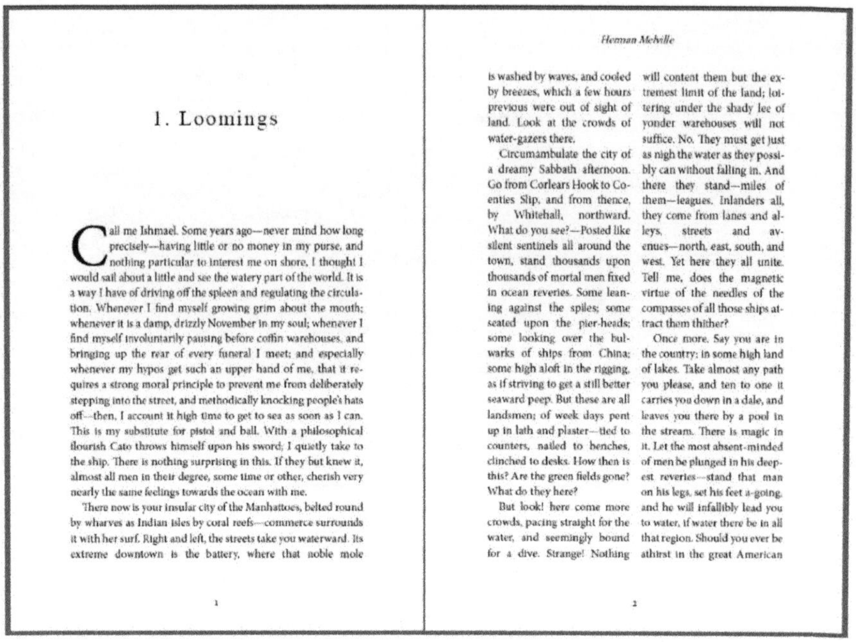

1. Loomings

Call me Ishmael. Some years ago—never mind how long precisely—having little or no money in my purse, and nothing particular to interest me on shore, I thought I would sail about a little and see the watery part of the world. It is a way I have of driving off the spleen and regulating the circulation. Whenever I find myself growing grim about the mouth; whenever it is a damp, drizzly November in my soul; whenever I find myself involuntarily pausing before coffin warehouses, and bringing up the rear of every funeral I meet; and especially whenever my hypos get such an upper hand of me, that it requires a strong moral principle to prevent me from deliberately stepping into the street, and methodically knocking people's hats off—then, I account it high time to get to sea as soon as I can. This is my substitute for pistol and ball. With a philosophical flourish Cato throws himself upon his sword; I quietly take to the ship. There is nothing surprising in this. If they but knew it, almost all men in their degree, some time or other, cherish very nearly the same feelings towards the ocean with me.

There now is your insular city of the Manhattoes, belted round by wharves as Indian isles by coral reefs—commerce surrounds it with her surf. Right and left, the streets take you waterward. Its extreme downtown is the battery, where that noble mole

1

Herman Melville

is washed by waves, and cooled by breezes, which a few hours previous were out of sight of land. Look at the crowds of water-gazers there.

Circumambulate the city of a dreamy Sabbath afternoon. Go from Corlears Hook to Coenties Slip, and from thence, by Whitehall, northward. What do you see?—Posted like silent sentinels all around the town, stand thousands upon thousands of mortal men fixed in ocean reveries. Some leaning against the spiles; some seated upon the pier-heads; some looking over the bulwarks of ships from China; some high aloft in the rigging, as if striving to get a still better seaward peep. But these are all landsmen; of week days pent up in lath and plaster—tied to counters, nailed to benches, clinched to desks. How then is this? Are the green fields gone? What do they here?

But look! here come more crowds, pacing straight for the water, and seemingly bound for a dive. Strange! Nothing will content them but the extremest limit of the land; loitering under the shady lee of yonder warehouses will not suffice. No. They must get just as nigh the water as they possibly can without falling in. And there they stand—miles of them—leagues. Inlanders all, they come from lanes and alleys, streets and avenues—north, east, south, and west. Yet here they all unite. Tell me, does the magnetic virtue of the needles of the compasses of all those ships attract them thither?

Once more. Say you are in the country; in some high land of lakes. Take almost any path you please, and ten to one it carries you down in a dale, and leaves you there by a pool in the stream. There is magic in it. Let the most absent-minded of men be plunged in his deepest reveries—stand that man on his legs, set his feet a-going, and he will infallibly lead you to water, if water there be in all that region. Should you ever be athirst in the great American

2

However, while by the middle of the 19th century, Ramean general textual maps have disappeared altogether, conceptual maps, which are better known today as conceptual models, concept maps as symbols of things in the world continued to evolve and became more and more non-verbal and specialized at the same time. An example is Mendeleev's periodic table in 1869.

While taxonomic maps as illustrated by Linnaeus's taxonomic tables and Mendeleev's periodic tables i (Figure 17) can be traced back to Ramean maps (Figures 18 and 19), other modern scientific conceptual maps find their origins back to the more logical maps such as those of Leonhard Euler John Venn, and Lewis Carroll. The best example of such maps, or if you prefer, visual conceptual models are the diagrams found in organic chemistry (Figure 18). In regard to the organic chemical diagram it should be noted that the symbol for a carbon compound is not only completely non-verbal, structural and spatial, a chemist fluent in this language, after looking at the structure in Figure 18 immediately knows that the "picture" contains the following information about the hydrocarbon compound.

Figure 17. Mendeleev's first periodic table, which consisted of six columns of elements and their atomic weight. Image source: http://www.rsc.org/education/teachers/resources/periodictable/pre16/develop/mendel4.jpg

			Ti = 50	Zr = 90	? = 180
			V = 51	Nb = 94	Ta = 182
			Cr = 52	Mo = 96	W = 186
			Mn = 55	Rh = 104,4	Pt = 197,4
			Fe = 56	Ru = 104,4	Ir = 198
			Ni = Co = 59	Pd = 106,6	Os = 199
H = 1			Cu = 63,4	Ag = 108	Hg = 200
	Be = 9,4	Mg = 24	Zn = 65,2	Cd = 112	
	B = 11	Al = 27,4	? = 68	Ur = 116	Au = 197?
	C = 12	Si = 28	? = 70	Sn = 118	
	N = 14	P = 31	As = 75	Sb = 122	Bi = 210?
	O = 16	S = 32	Se = 79,4	Te = 128?	
	F = 19	Cl = 35,5	Br = 80	J = 127	
Li = 7	Na = 23	K = 39	Rb = 85,4	Cs = 133	Tl = 204
		Ca = 40	Sr = 87,6	Ba = 137	Pb = 207
		? = 45	Ce = 92		
		?Er = 56	La = 94		
		?Yt = 60	Di = 95		
		?In = 75,6	Th = 118?		

Figure 18. While the periodic table of elements (left), which is a higher order concept map can be traced back to Ramus's diagrams, the structural largely non-verbal symbols in organic chemistry (right) can be traced back to the logic diagrams of Euler, Venn and Carrol

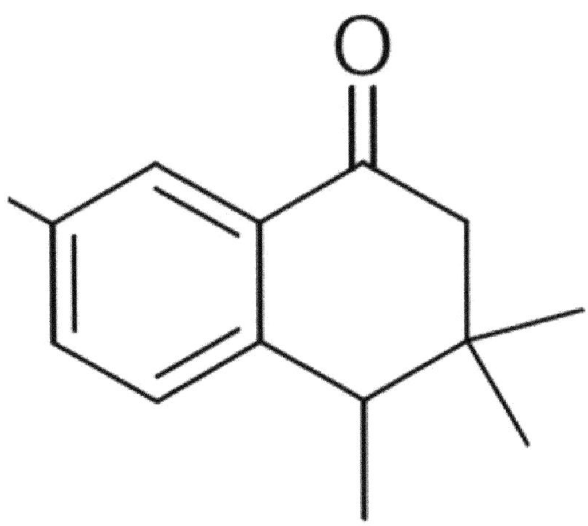

Figure 19 is the same hydrocarbon in the previous figure, only labelled to bring in the verbal element for teaching purposes, which is fine for teaching, but is redundant for an advance researcher calculating organic chemistry equations. It should also be noted that all of the above are concept maps for very specific domains. The concept of concept maps as meta-cognitive tools is the subject for the next chapter.

Figure 19. Hydrocarbon compound labelled with words, making it more understandable to students and novices

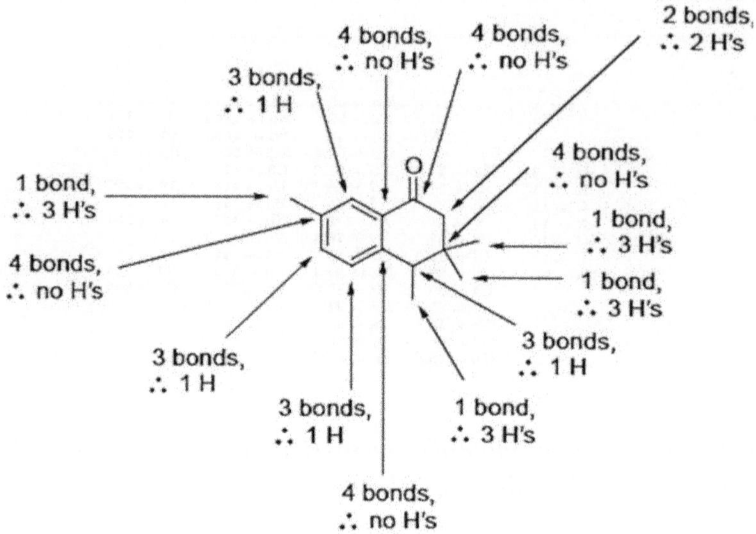

CONCLUSION

At this juncture we have shown that in their evolution, conceptual maps became more and more abstract until Ramean maps as general textual maps disappeared altogether and other maps such as Venn diagrams became not only more abstract but also more specific as illustrated in the foregoing examples of the organic chemistry conceptual models. Ramean concept maps mostly disappeared into the text book partly because of a technical issue as it was more efficient and cost effective to lay out the simple logic behind the maps in a tabular form rather in the original vertical arrangement separated by brackets. Moreover, the logic behind the maps, while exiting and novel in the 16th century, became trivial in the 19th.

In addition, concept mapping after the Renaissance up to early 20th century followed a pattern that began at a time when the use of diagrams was prevalent throughout education thanks largely in part because of Ramus and the Protestant movement. That movement took the printing press and the idea of mapping out information in hierarchical diagrams and spread the book culture along with Ramean analysis throughout Europe, laying the ground work for the popularization of the accepted "Order of Things" during and after the Scientific Revolution. As the scientific age continued, however, Ramean analysis and the logic associated with it had been so inculcated in the Western Mind, that students no longer needed diagrams as an aid to textual analysis and exegesis.

The precious chapters on the history of conceptual maps also showed that these maps can be divided into more or less verbal maps as represented by the Ramean diagrams to essentially mathematical maps, as represented by the moderns of the previous chapter, Leonhard Euler, Lewis Carroll, and John Venn. These maps were not only highly logical and mathematical but also accessible to only a select and esoteric few, the logicians and mathematicians. During roughly the same time period, Ramean maps also became more specialized and restricted in their domain appearing in different forms in the natural sciences.

The modern logical/mathematical maps beginning with Euler, Venn and Carroll and continuing into the present century, claimed to be the universal meta-language envisioned by Leibniz when he was searching for a Universal Language and a Universal Character (i.e. notational symbol system). In this context, Charles Sanders Peirce significance is that he provided a sound foundation for concept mapping through his forays into semiotics as a science. He provided us with a theory, his triadic model of the relation between sign and symbol, interpreter and referent or object that framed research as to why and how cognitive mapping works. The problem with Peirce's and his fellow logicians, however, is their conceptual languages, is they could, and it is still the case, be understood, much less mastered only by only a few. It was to Lewis Carroll's credit that he, as a university instructor, at least tried to bring highly abstract logical concepts down to the masses to his undergraduate students, not only through his conceptual diagrams as instructional devices but through mathematical games. Carroll/Dodson actually tried to make the learning of an extremely difficult subject fun, something unheard of in his day.

REFERENCES

Anellis, I. H. (1995). Peirce rustled, Russell pierced: How Charles Peirce and Bertrand Russell viewed each other's work in logic, and an assessment of Russell's accuracy and role in the historiography of logic. *Modern Logic*, *5*, 270–328.

Barker, P., & Goldstein, B. R. (2001). Theological foundations of Keplers astronomy. *Osiris*, *16*, 88–113. doi:10.1086/649340

Bricken, W. (2016). *Iconic math*. Available at http://iconicmath.com/logic/boundary/

Brown, G. S. (1969). *Laws of form*. Unwin and Allen.

Burch, R. (2014). Charles Sanders Peirce. *Stanford encyclopedia of philosophy*. Available at http://plato.stanford.edu/entries/peirce

Casper, M. (1959). *Kepler*. Collier Books.

Cohen, M. (1996). *Lewis Carroll: A biography*. Vintage Books.

De Chiara, , Erra, & Scarano. (2003). VENNFS: A Venn-diagram file manager. *Proc. IEEE Information Visualization (IV03)*, 120-126.

Debnath, L. (2009). The legacy of Leonhard Euler–a tricentennial tribute. *International Journal of Mathematical Education in Science and Technology*, *40*(3), 353–388. doi:10.1080/00207390802642237

Diagrams, E. (2004). *Conference. Brighton, UK*. Available at https://www.cs.kent.ac.uk/events/conf/2004/euler/eulerdiagrams.html

Edwards, A. W. F. (2004). *Cogwheels of the mind: The story of Venn diagrams*. JHU Press.

Finkel, B. F. (1897). Biography: Leonard Euler. *The American Mathematical Monthly*, *4*(12), 297–302. doi:10.2307/2968971

Fisch, M. H. (1986). *Peirce, semeiotic, and pragmatism: Essays*. Indiana University Press.

Fish, A., Flower, J., & Howse, J. (2005). The semantics of augmented constraint diagrams. *Journal of Visual Languages and Computing*, *16*(6), 541–573. doi:10.1016/j.jvlc.2005.03.001

Hammer, E. M. (1998). Semantics for existential graphs. *Journal of Philosophical Logic*, *27*(5), 489–503. doi:10.1023/A:1017908108789

Hoopes, J. (Ed.). (2014). *Peirce on signs: Writings on semiotic by Charles Sanders Peirce*. UNC Press Books.

Howse, J., Molina, F., Taylor, J., Kent, S., & Gil, J. (2001). Spider diagrams: A diagrammatic reasoning system. *Journal of Visual Languages and Computing*, *12*(3), 299–324. doi:10.1006/jvlc.2000.0210

James, W. (1975). *Pragmatism* (Vol. 1). Harvard University Press.

Kankkunen, M. (2001). Concept mapping and Peirce's semiotic paradigm meet in the classroom environment. *Learning Environments Research*, *4*(3), 287–324. doi:10.1023/A:1014438905784

Kankkunen, M. (2004). How to acquire "The Habit of Changing Habits": The marriage of Charles Pierce's semiotic paradigm and concept mapping. In *Concept Maps: Theory, Methodology, Technology Proc. of the First Int. Conference on Concept Mapping*.

Koestler, A., & Butterfield, H. (1959). *The sleepwalkers: A history of man's changing vision of the universe*. The Universal Library.

Lovejoy, A. O. (2011). *The great chain of being: A study of the history of an idea*. Transaction Publishers.

Malinowski, B., Crookshank, F. G., Ogden, C. K., & Richards, I. A. (1923). *The meaning of meaning; a study of the influence of language upon thought and of the science of symbolism*. Kegan Paul, Trench, Trubner.

McLuhan, M., & McLuhan, E. (1992). *Laws of media: The new science*. University of Toronto Press.

Moore, E. C., & Robin, R. S. (Eds.). (1964). *Studies in the philosophy of Charles Sanders Peirce, Second Series*. U. of Massachusetts Press.

Peirce, C. S. (1931). Existential Graphs. In Collected papers of Charles Sanders Peirce. Volume 4, Book 2 (pp. 347–584). Harvard University Press.

Simon, G. (1975). 8.3. Keplers astrology: The direction of a reform. *Vistas in Astronomy*, *18*, 439–448. doi:10.1016/0083-6656(75)90122-1

Tillyard, E. M. W. (2011). *The Elizabethan world picture*. Transaction publishers.

Whitehead, A. N., & Russell, B. (1912). *Principia mathematica* (Vol. 2). University Press.

Wilson, R. (2008). *Lewis Carroll in numberland: His fantastical mathematical logical life*. Allen Lane.

Wolfram, S. (2016). *Demonstration project: Lewis Carroll's bilateral diagram*. Available at http://demonstrations.wolfram.com/LewisCarrollsBilateralDiagram

Section 2

The Retrieval of Conceptual Maps as Learning Tools

Section 2 analyzes and discusses the construction of concept maps as educational tools, retrieved from the distant past of the logical semantic diagrams used by Peter Ramus and his students to teach student the content of the textbooks in his day. The first of these chapters covers the work of Albert Upton, who taught English not as literature appreciation but as analytical tool to better understand literature followed by the work of David Ausubel, an educator, Professor Ausubel was a advocate of a teaching philosophy called constructivism which viewed learning as a student centered process where students are guided by their teachers to discover meaning in the lessons they encounter. J.D. Novak, another educator, will take Ausubel's ideas as a framework and re-invent concept mapping as we understand the concept today. Chapter 4 gives a classification of concept maps used today and discusses the work of Tony Buzan and David Hyerle along with a further analysis of Novak's maps.

Chapter 3
Verbal Concept Maps

ABSTRACT

Conceptual maps have become so advanced, and specific to narrow fields that today only specialized experts such as mathematicians and computer scientists are able to understand them. The esoteric logical and mathematical diagrams such as Euler's and Venn's specialized maps were nearly invisible because they were never very visible in the first place. Concurrently, general verbal and logical conceptual diagrams became nearly invisible because they were internalized as structures as verbal maps as figure receded into the ground as the accepted paradigm of logical thinking and exegesis among the educated. This chapter covers the retrieval of Verbal Concept mapping traceable to the works of Peter Ramus, focusing on the works of Albert Upton, David Ausubel, and J.D. Novak.

INTRODUCTION

The following classification chart summarizes where this book has been and where it is going.

Figure 1 is a working hypothesis in map form of concept maps according to their creator covered in this book so far and to be covered in this chapter. According to this hypothesis, conceptual models can be divided into two kinds, visual models of the world and models of the internal operations of the mind. In the classification chart as hypothesis we have divided models of the world into spiritual and scientific with Ramon Lull constructing spiritual maps that were more spiritual while Johann Kepler constructed conceptual

DOI: 10.4018/978-1-5225-2184-6.ch003

Figure 1. A classification chart of the persons and their conceptual models covered so far and to be covered

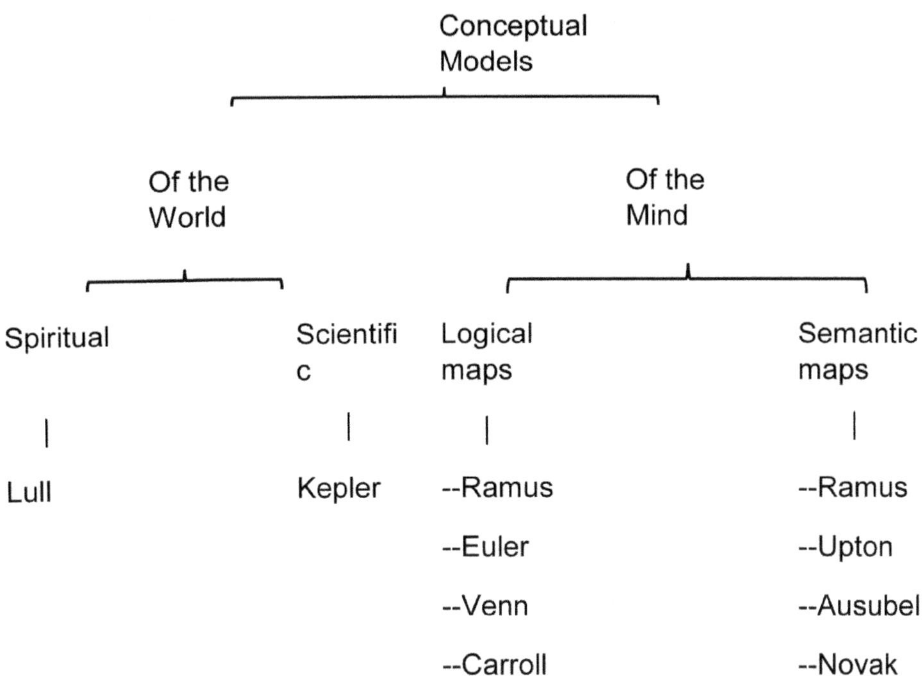

models that mere more scientific. At the same time, while we attributed concept mapping to both figures, it would be more accurate to call Ramon Lull's maps, "spiritual diagrams" and Kepler's conceptual maps "geometric models." This is because, the term "concept maps" was coined by J.D. Novak, the educator to be discussed later in the chapter.

Under the family of "conceptual models of the mind" the classification chart separates logical maps from semantic maps, with "concept maps" in the narrow sense of the term as defined by Novak, evolving out of this category of conceptual models. Petrus/Peter Ramus's maps, in the chart, have been defined, according to the chart, as both logical and semantic maps, as they were considered to be both in his day. However, if we were to take a strictly linear approach to the evolution of concept mapping, it is argued that concept maps, will be defined by Novak, evolve strictly out of Ramus's maps as semantic maps. Moreover, the logic maps of Euler, Venn and Carroll, will evolve in other direction that is not only more logical but also more mathematical ending in their most recent and significant manifestation as concepts maps in computing, a development that is not covered in this book, but a most valuable area of research that needs further investigation. However, to summarize, the development of Upton's, Ausubel's and Novak's semantic maps can be directed traced back to the semantic aspects of Ramean maps in the 16th century discussed in an earlier chapter.

Albert Upton

In the twentieth century the concept map in its verbal and Ramean form was retrieved as an educational device by a relatively unknown professor at a private liberal arts Quaker college in California best known for its most famous alumnus, former President Richard Nixon. The name of the professor was Albert Upton, who graduated from the University of California, Berkeley, with a Ph.D. in English, and later a professor in the same subject. Professor Upton, however, was not the typical English professor of the time in the early 50's and 60's. Albert Upton was a polyglot who may have had an immeasurable IQ in that he was said to have gotten perfect scores on several IQ tests that he took, including versions of the Stanford Benet IQ test and the entirely non-verbal Raven Progressive Matrices tests. Professor Upton was also out of the ordinary during his time in that, contrary to prevalent psychological opinion, he believed that intelligence, even as measured by IQ tests, was malleable and therefore could be taught (Raven, 2002; Costa, 2008).

He also taught Russian and had a keen interest in semantics, before and after studying with the famous, I.A. Richards, who co-authored one of the best known linguistics books of the 20th century with C. K. Ogden, a book called, the *Meaning of Meaning*. (Malinowski, 1923). From I.A. Richards, he learned about Basic English, Ogden's and Richard's attempt at converting English to a world language and lingua franca by reducing the language to a set of 800 words and their derivatives. As a controlled and simplified language, BASIC, which stands for "British American Science Industry and Commercial English, consisted of only 18 verbs among 100 "operators," which included articles and prepositions and a set of 600 nouns divided into 400 general words and 200 which could be pictures of things. The root set of 800 words was rounded out by 150 qualities and their opposites (Richards, 1943).

While he was at Whittier College, Albert Upton, who became chair of the English department, changed the format and content of the college's two required Freshman English courses, designated at most colleges and university with such rubrics as English 101 and English 102. That change was to change the courses from traditional English literature reading and composition courses where English was the subject to semantic courses where English was taught as a tool for understanding and thinking more clearly, critically and effectively.

Unlike most of his fellow college English teachers, Professor Upton approached the subject as a positivist scientist steeped in practical knowledge of foreign languages and trained through his association with I.A. Richards with one of the most brilliant scholars of English at that time. Consequently, when he taught *Hamlet* in the first semester English course, he made his students translate parts of into BASIC so they could understanding the meaning of what they were reading before they started considering Shakespeare's emotional and aesthetic impact. In addition to analyzing meaning through translating archaic English to simpler analytical BASIC, Upton (1941) had students construct concept maps of different passages from the play. Figure 2 is a scan from his textbook at that time, *Design for thinking: A first book in semantics.* As can be seen from the textbook scan, the diagram bears a striking resemblance to the Ramean diagrams discussed in the previous

Figure 2. A scan of a page from Upton's "English" textbook using semantic maps (Upton, 1941)

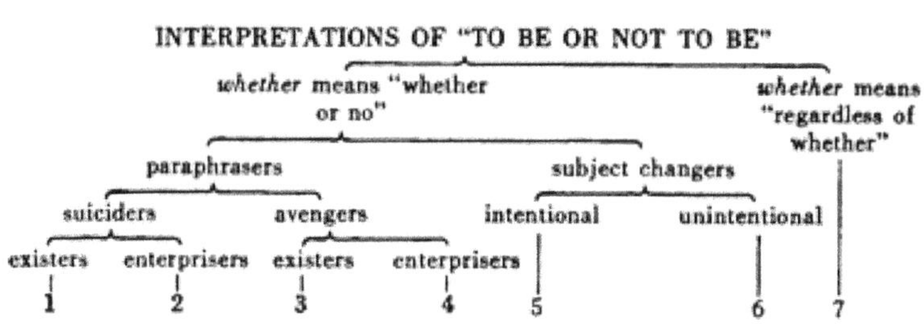

Interpretation by	*Interpretation of "to be or not to be"*	*Interpretation of "whether 'tis nobler to suffer . . . or to take arms"*
PARAPHRASERS		
Suiciders		
1. Existers	(1) To live (2) or not to live	(1) To suffer troubles (2) or to end troubles by suicide
2. Enterprisers	(1) Is this enterprise of suicide to take place (2) or not to take place	(2) To suffer troubles (1) or to end troubles by suicide

chapter, brackets and all, only the brackets are horizontal and the concepts are to be read from top down instead of left to right.

In this regard, Professor Upton could be regarded not only to have retrieved the meme of concept mapping but also as a pioneer of the idea of critical thinking, something that is much in vogue among educators today, because the main purpose of the maps and Upton's teaching philosophy was to teach English as an instrument of critical and higher order thinking with the maps as ancillary aids.

Subsequently, the logician in Upton reduced all thinking processes, including the process of understanding what is read, to four logical statements or questions:

- **Classification:** What is this a kind of and what are the kinds of this?
- **Structural Analysis:** What is this a part of and what are the parts of this?
- **Qualification:** What is this a quality of and what are the qualities of this?
- **Operational Analysis:** What is this a stage of and what are the stages of this?
- **Analogical Transfer:** What is this an analog of and what are the analogs of this?

All questions which Upton thought could be answered through the construction of four or five different kinds of maps which could what can be called semantic maps.

For example, in figure 3, we have two instances of Professor Upton's early semantic maps. The classification chart to the left answers the question, "What are plants and animals kinds of and what are different kinds of organisms?"

Figure 3. Two Upton semantic maps laying out a classification of organisms (left) and the parts of the body (Upton, 1941)

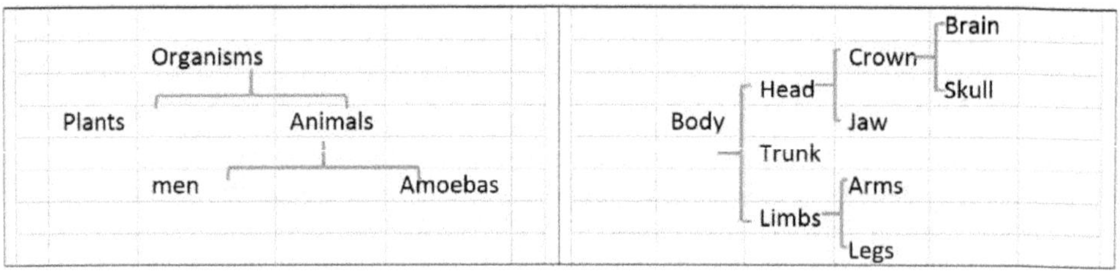

and the second map (right), "What are crown, jaw, brain, etc. parts of and what are the parts of the body?"

Another map that he had his students construct was an operational analysis map, which is simply an ordered list. Yet another map, the analogical map, is shown to the right in figure four, by which a student can map out the meaning of Psalm 23-28, which begins, "The Lord is my shepherd. I shall not want," with an analogical map shown in figure 4 which for the mathematically minded, A: B:: C: D, while the more verbal person will immediately see that " as the shepherd tends his sheep, the Lord will take care of his followers.

In his English classes, Upton – the semanticist – would make his students work on thinking lessons based on this theory much like instructors of earlier grades had their students do grammar lessons. An example of such an exercise is seen in figure 5, which was to have students understand the semantic principle of words as signs pointing to things or other signs such as pictures as in in the following exercise, where the student is to write the word that applies to the class of things pictured at the top, and then select the images that belongs to that class.

The verbal answer to question 7, of course is "bird" and "B" as the letter to be circled corresponding to correct word. In this exercise, which consisted of 50 such items, students learn by practice, much like doing math problems not only the understanding that words point to things or other signs such as pictures, but that words and sentences have a logical syntax. In question 8, becomes apparent that the term "shoe" refers to a class of things related to foot ware.

Professor Upton has retrieved Ramus's conceptual maps as a general analytical tool, a meta-tool that can be used to increase understanding of subjects reading from Biblical scripture to patterns of probability. At the same time, while Albert Upton was developing a general tool for understanding partially based on graphical map analysis and construction, David Ausubel on the east coast of the United States was building a theoretical foundation based on constructivist learning theory to explain and frame conceptual mapping as a general organizational especially suited for education purposes.

David Ausubel

David Ausubel was born in 1918 and grew up in Brooklyn and studied at the University of Pennsylvania where he graduated with honors in 1939, receiving a bachelor's degree majoring in Psychology as a pre medical student, graduating

Figure 4. Upton's analogical map on the meaning of the Psalm 23-28 (Upton, 1941)

"THE LORD IS MY SHEPHERD"

Analogical universe (husbandry) = Contextual universe (Religion)

A) The shepherd — Tends — B) His sheep

C) The Lord — Encourages — D) His followers

Relevant relations: promotes the welfare of

Figure 5. Two questions from an Uptonian exercise (Tang, 2017)

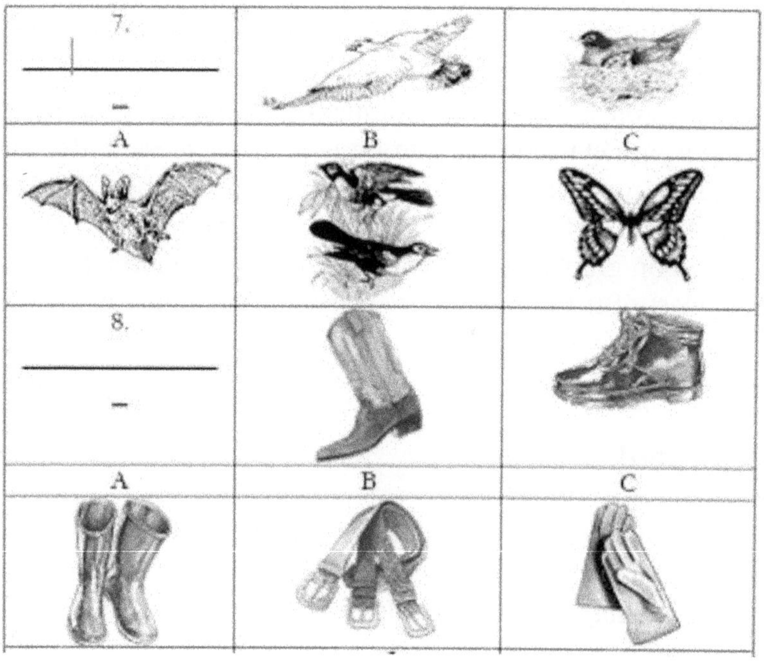

with a medical degree from Middlesex University after an internship at Gouverneur Hospital, located in the lower east side of Manhattan, New York, he went on to practice psychiatry with the US Public Health Service, where he became disappointed with psychiatry as a profession. Disillusioned with psychiatry, Dr. Ausubel returned to school and earned a M.A. and Ph.D. in developmental psychology from Columbia University in 1950 (Ivie, 1998)

In 1950 Ausubel accepted a position with the Bureau of Educational Research at the University of Illinois and remained with the Bureau for the next sixteen years, a time during which he published extensively on cognitive psychology. Ausubel left the University of Illinois in 1966 to accept a position with the Department of Applied Psychology at the Ontario Institute of Studies in Education in Toronto, Canada where he stayed for two years before becoming Professor and Head of the Department of Educational Psychology, Graduate School and University Center, City University of New York, which today is better known as simply CUNY. In 1976, he received the Thorndike Award from the American Psychological Association for "Distinguished Psychological Contributions to Education (Ausubel, 2016). As in the case of Albert Upton, Dr. Ausubel was concerned with how people acquire knowledge or how people think, especially as to what is today as higher order thinking. Unlike Upton, however, Ausubel arrived at the idea of using concept maps, or what he called graphic organizers as an aid to higher order thinking not from empirical practice in the classroom or from linguistic influences but from developmental psychology, especially Piaget's theory of cognitive development (Ivie,1998).

Jean Piaget's theory of cognitive development is a constructivist theory about the nature and development of human intelligence that deals with the nature of knowledge itself and how humans gradually come to acquire, construct and use that knowledge. This developmental theory of cognition postulated that how we think involves a progressive reorganization of mental processes resulting from biological maturation and environmental experience. Key to this cognitive concept is the idea that we construct an understanding of the world around us through our experience in everyday life during which time we encounter anomalies, or experiences that do not fit our previous constructs. It is in solving these anomalies, by assimilating them into our previous constructs and thereby reconstructing our knowledge paradigm that humans progress intellectually. This approach to learning was the foundation of an education theory that came be known as constructivism.

Constructivism

Educational constructivism is a philosophical viewpoint about the nature of knowledge and how that knowledge is acquired with an emphasis on learning through the interaction between the learners' experiences and their ideas, acknowledging their uniqueness and complexity by encouraging, utilizing and rewarding those characteristics as integral to the learning process (Wertsch 1997). Salient aspects of that student centered approach or philosophy include:

The importance of the background and culture of the learner whereby learners arrive at their own versions of the truth as influenced by their worldview with an emphasis on their symbol systems, such as language, logic, and mathematical as conditioned by their society. Background and cultural importance involves starting with the students' intellectual background and building on background progressing in stages from lower to higher levels of thought and understanding,(Piaget, 1969; Vygotsky, 1978).

Figure 6. An Ausubel graphic organizer. Image source: https://hlkemp.files.wordpress.com/2014/02/graphic-organizer-topic.jpg

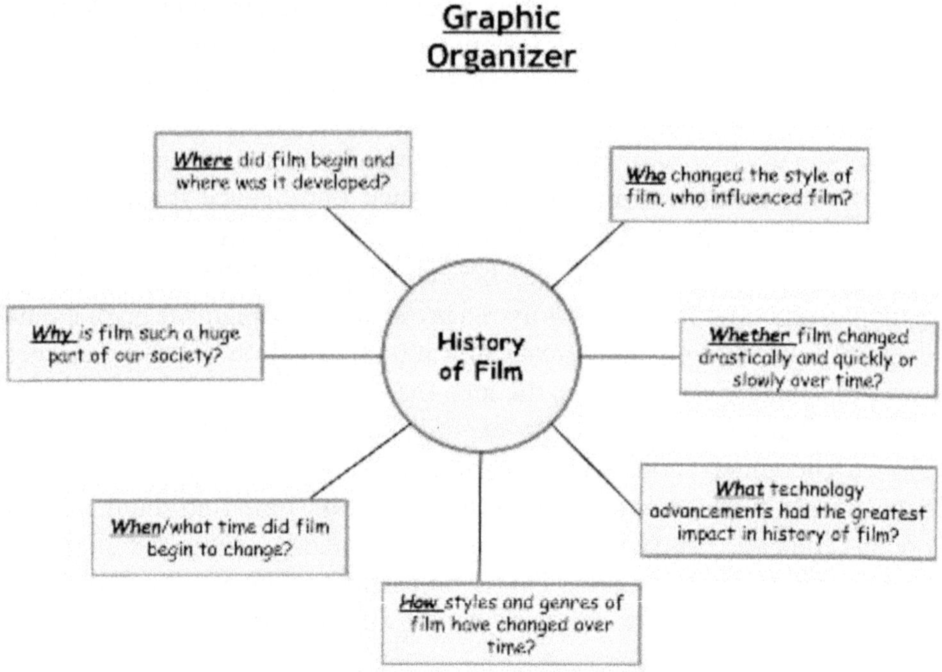

Responsibility for learning rests primarily on the student who must be actively involved in the learning process, unlike previous educational approaches where the responsibility rested with the instructor to teach and where the learner played a passive, receptive role. In this active role, students construct their own knowledge with the teacher acting as a guide to help them find meaning in the knowledge they acquire vis a vis the external world where they have to operate outside the classroom (Von Glasersfeld, 1995).

The motivation for learning can be increased by related what has been learned with what will be learned in stages which sustains the motivation to advance intellectually through instructor feedback based on the students' inner development and advancement that increases confidence in the students' ability and potential to acquire higher and newer forms of knowledge. to ==earner's confidence in his or her potential for learning where learners are challenged within close proximity to, yet slightly above, their current level of development. By experiencing the successful completion of challenging tasks, learners gain confidence and motivation to embark on more complex challenges. (Prawat, 1994; Vygotsky 1978)

Ausubel's greatest contribution to the development of conceptual mapping is to tie a theory of mental mapping to the above three aspects of constructivism by emphasizing that these maps can be used to organize the mental constructs that Piaget called schemas. Ausubel defined such schemas as a set of linked mental representations of the world that are used to understand and to respond to that world. To use a modern analogy, schemas are like a personal operating system equipped with flow charts which frame and manage the way we perceive, experience and learn from our interaction with the external world.

Applying Piaget's developmental psychology and constructivism directly to the classroom, Ausubel argued that a better way to teach and learn was not to pour information directly into the student brain as a tabula rasa, usually via the lecture, send them home to do read the assigned textbook on which the lectures are based and do homework which usually meant more reading and working up assigned problems and then test them on the subject covered. A better way to teach was to give students advance schema, or what he called organizers to connect them with pre-existing student schemas and link them together in a way that can help that understand the new knowledge and content that is needed to advance academically and intellectually. (Ivie, 1998).

In Ausubel's pedagogical universe, advance organizers come in two types, comparative organizers and expository organizers. Comparative organizers are supposed to activate existing schemas already existing as mental structures in the mind of the student and act as a mnemonic catalyst to stimulate active

Figure 7. Comparative advance organizer (left) where instructor feeds organize to students memory bank (center) for them construct a meaningful mental map (right)

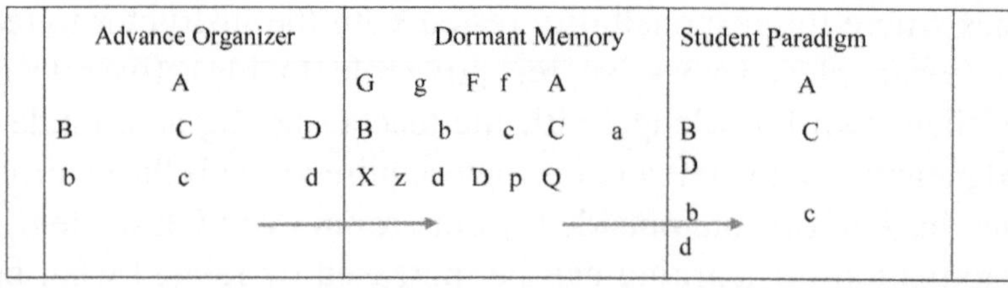

working memory to be conscious of what was previously dormant in memory. By acting as reminders, the advance organizers, consisting of key knowledge nodes or anchors, change a student's internal knowledge structure, after their presentation, by having students try to match their previous knowledge with the new knowledge presented as advance organizers. Comparative advance organizers are used both to integrate as well as discriminate or distinguishing concepts by connecting new ideas with similar concepts and selecting out those concepts which do not fit into the new mental paradigm being created (Ausubel, 1968; Woolfolk 2010).

As per figure 3, the instructor would feed the comparative advance organizer into the students' memory bank vis a vis a lecture, homework, or as a problem, which activates the student's mind to filter and sort out stored data to construct a mental conceptual map that matches the original organizer input. Instructors would employ an expository advance organizer, on the other hand, to feed in new information to activate the experiential working memory to search for an analogous map based on the students' mental structures based on their actual experience outside the classroom.

Both comparative and expository organizers are based on Ausubel's assimilation theory that states that students learn best when they can build on knowledge already acquired. Comparative advance organizers connect existing schemas with the databank that stores intellectual concepts to reorganize existing dormant nodes in the working memory as a mental map to match the instructor's template. Their function is to enable students to construct a new internal schema by making new connections and filtering out memory units that do not fit or are unnecessary to the new mental paradigm that has been assimilated in a lecture or lesson.

On the other hand, expository organizers provide students with new knowledge that will be needed to understand the upcoming information and concepts (Woolfolk, 2010), based on overcoming the students' innate fear

Figure 8. Analogical map "laying out" the correspondence between the Big Bang Theory and Genesis. Note the similarity between Ausubel's map and Upton's analogical map

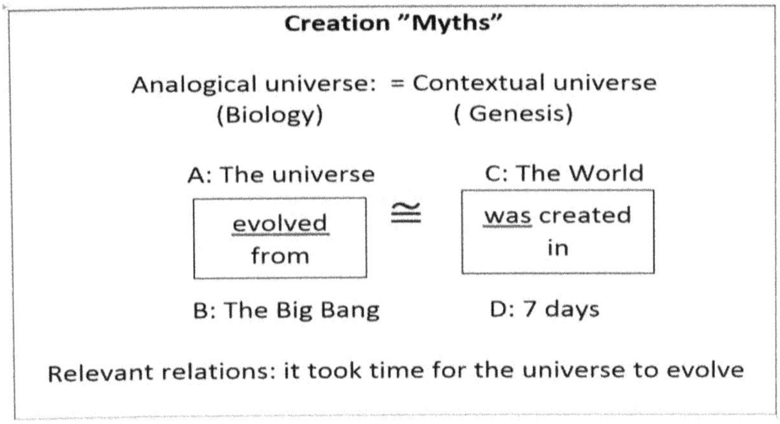

of and resistance to the new. Expository organizers do this by connecting student's former knowledge with the new knowledge in a way that makes the new knowledge more plausible in the students' minds. Both comparative and expository organizers are based on Ausubel's assimilation theory argues that the best way for students to acquire new knowledge is to build on knowledge previously learned and stored in the memory banks (Mayer, 1979). Comparative organizers rearrange acquired unsorted information into new frames with new connections between the nodes, or what Ausubel called "anchors," while expository organizers link old frames of schema with new schema by the way of analogical congruency (Ausubel, 1968).

For example, the Darwinian theory of evolution may not only be new to students but some may react to the theory not only with unconscious fear, but with denial, perhaps because of their religious background. Expository advance organizers are designed to make new knowledge more plausible and less fearful by positively connecting the new knowledge to generally relevant knowledge that is the instructor believes to be already present in the students. In figure 5 we see how an instructor might introduce an expository organizer to the Darwinian theory by drawing the parallel between Genesis seven days of creation with scientific evolution, by arguing that the "seven days" were equivalent to 2 billion years, making the age of universe, according to the Bible, about 14 billion years, close to the scientific estimate.

Both comparative and expository organizers serve as what Ausubel call "anchors" or fundamental structures that hold together and connect either to existing mental structures (comparative organizers) or to new knowledge (expository organizers) (Ausubel, 1968). Dr. Ausubel's cognitive schema theory laid out the psychological foundation for a theory of conceptual maps which another developmental constructivist psychologist, Robert Novak would reify in what Novak would call "concept maps."

J.D. Novak

The single person most responsible for concept mapping as it is understood today was J. D. Novak, a constructivist education scholar, who took Ausubel's *mental* organizers and putting them in visual form by actually drawing them up on paper and calling the drawings, "concept maps." This bold and in retrospect simple step occurred when he and his colleagues at Cornell were trying to keep track of and understand changes in children's knowledge of science in a less reductionist and linear fashion. This break from the traditional behavioral psychological epistemology, training and methods, Novak (2010) attributes to his discovery of the works of Ausubel.

As a graduate student at the University of Minnesota in the 1950's, the only learning psychology taught was behavioral psychology, and this largely sought to remove meaning from learning experiments, either by using animals, nonsense syllables, paired-word associations, or other material that would not "contaminate" experiments due to the almost idiosyncratic nature of a learner's meanings of concepts or ideas. I thought then that behavioral psychology had essentially no relevance to human learning of the kind I was interested in. Fortunately for our research group, Ausubel's assimilation theory of meaningful learning was published in 1963, and this provided a solid theoretical foundation for the work we were interested in doing (21).

This assimilative theory of learning should be contrasted to the reductionist behaviorist theory of learning which emphasized recall and rote memory.

The key idea in Ausubel's theory is the distinction between learning by rote versus learning meaningfully. When learning by rote, no effort is made to relate new ideas with relevant existing ideas in cognitive structure. In meaningful learning, the learner chooses to integrate in some substantive way new concepts and propositions with existing relevant ideas in her/his cognitive structure (21-22).

J. D. Novak developed his technique of using mapping as graphic organizers during a period of research on students when he and his colleagues at Cornell decided to apply Ausubel's theories by tracing their own understand of changes in children's knowledge of science in a less linear and more holistic and constructivist fashion. The first concept maps, therefore, were used to keep track of the team's knowledge about student progress in regard to a research project they were conducting. Afterwards, Novak and his team extended their mapping as diary function to the idea of concept maps in general as tools that they believed were isomorphic with any mental schemata. The distinguishing characteristics of general concept maps as meta-cognitive tools according to Novak are as follows: (Novak, 2010):

First, concept mapping is grounded in Ausubel's Assimilation Theory that posits that new knowledge can be learned most effectively by relating it to previously existing knowledge. Concept Maps may be viewed as a methodological tool of Assimilation theory that visually displays fundamental elements of the theory.

Secondly, Novak's concept maps are organized semi-hierarchically. The basic reason behind the hierarchical arrangement of concepts in concept maps comes from Ausubel's notion of subsumption, that more general, superordinate concepts subsume more specific, detailed concepts. At the same name Novak concedes that concept maps per se, need not be always hierarchically structured.

Labeled Links. Hierarchical and non-hierarchical relations are defined by the use of linking phrases between concepts (Novak, 1984) to form a meaningful proposition, which is a basic unit of knowledge according Concept mapping theory does not constrain the labels that can be used, allowing map makers more freedom and precision in describing the relationships among concepts.

Given these general characteristics, concept maps can be defined in detail through a concept map such as the one found in figure 9 (Novak, 2008).

Basically a concept map is a structure of nodes, which contain things and concepts that are interconnected with labelled arrows, and because of its general use in education, the question is not whether or not the maps are being effectively used in the classroom but in which classrooms. Through data mining bibliometric techniques, the authors found the following the number of articles concerning the effective use of concept mapping in

Figure 9. A concept map defining "concept maps" (Novak, 2008)

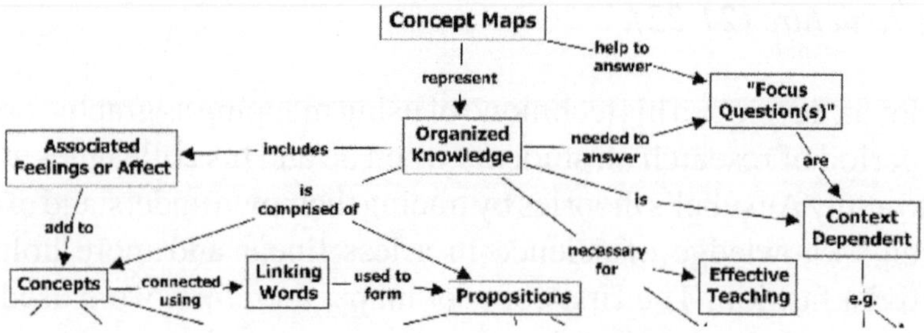

Figure 10. Number of Google Scholar articles on using concept mapping in different fields on the undergraduate level

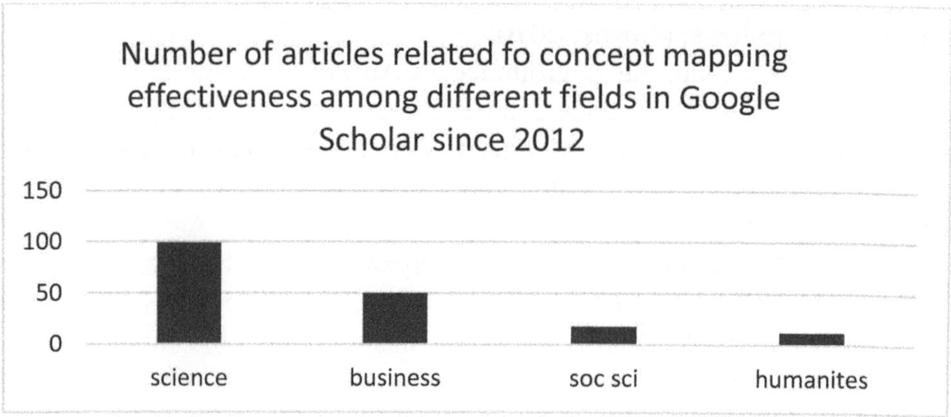

the undergraduate fields of science, business, the social sciences and the humanities since the year 1994.

In regard to the different fields within the sciences, where concept maps appear to be used most, a similar data analysis technique found that the number of articles regarding their effectiveness were more or less evenly distributed among engineering, nursing, biology, chemistry and physics disciplines. In another bibliometric analysis of concept maps in STEM education, the authors found that concept mapping's highest application after science per se was in its field of origins, education, from which the technique appears to have spread to business, computer science, science and the social sciences in that order (Figure 11). It might also be noted that the analysis on the growth of concept mapping in computer science shows a linear progression exceeding that of its use in the sciences --- mainly physics, chemistry and biology --- in

Figure 11. A bibliographic analysis of concept mapping applied to different fields, showing increase in all and most used in business and education

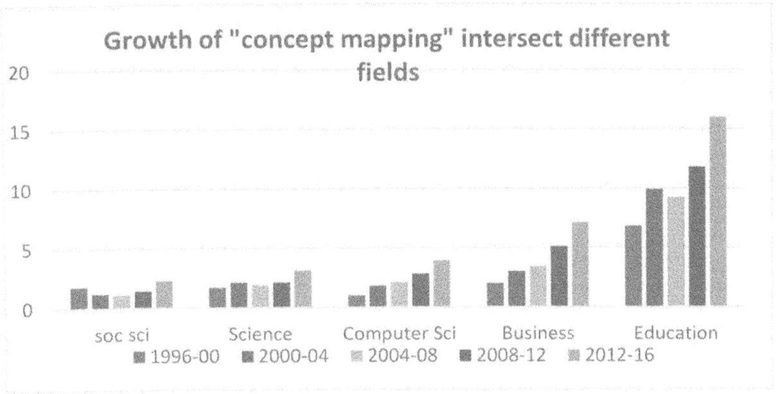

2016. This is all the more remarkable in that computer science as a field that is much smaller that the set of "science", delineated in the search analysis, confirming the correlation between the numbers in Figure 10 and 11.

The above bibliometric analyses indicate not only that concept mapping is an effective meta-cognitive technique for organizing and processing information, but also where the technique has been used most: education, the sciences, and science education. This macro analysis parallels L. H. Clayton's study, who, after analyzing several different concept mapping studies in greater detail concluded that 1) concept mapping in general results in generally positive effects on academic performance; 2) that the technique improves student thinking ability; and 3) concept maps can be used effectively in teaching; and another reason for the effectiveness, may a motivation factor (Clayton, 2006). The following are only a few examples of the techniques demonstrated effectiveness.

W. M. Able and M. Freeze's study that evaluated concept mapping as a clinical teaching-learning activity to measure critical thinking through identification of nonlinear relationships among different components of the nursing process. Their experimental sample involved 28 associate degree nursing students who completed a total of four concept maps as integral assignments in their courses during a period of three semesters. The investigators of this research project found that as the students progressed through the curriculum, they steadily increased their mean scores when evaluated with a nonlinear cross-linked measuring tool (Abel, 2006).

Figure 12. A systems thinking model that is isomorphic to a concept map. Image source: http://www.mdpi.com/sustainability/sustainability-08-00005/article_deploy/html/images/sustainability-08-00005-g002-1024.png

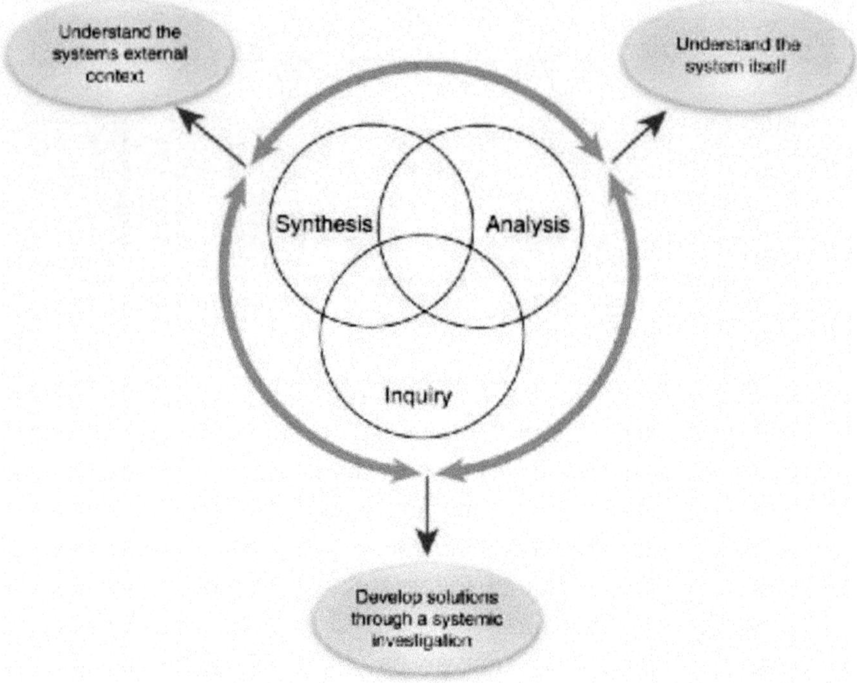

In another experimental study also conducted by nursing education researchers, the investigators designed a pretest-posttest experiment to evaluate the effectiveness of concept mapping in developing critical thinking skills in baccalaureate nursing students with an n of 76, divided into two groups for control. The experimental group (n 44) was taught to use concept mapping of patient information to prepare for clinical experiences while the control group (n 32) was taught in the traditional manner with the same course work. Critical thinking skills were measured with the California Critical Thinking Skills Test which resulted in the experimental-group scores improved significantly ($p < .05$) on the overall score while the control-group scores improved only on the evaluation subscale and declined significantly on the inference subscale, suggesting that that concept mapping is effective in helping students develop critical thinking skills (Wheeler, 2003).

Furthermore, this chapter's research showed that Novak's maps were not only congruent with constructivist learning theories but also with systems thinking. In a study conducted by researchers at the Ben Gurion University in Israel on Israeli high school system, it was found that complex systems like those that make up the human body have become increasingly important as a focus of high school education. When the researchers used Novak's maps to teach both the content and structure of the systems such as the human body they found significant effectiveness in using concept maps to improve systems thinking. As per figure 12, systems thinking involves the ability to make inquiries, synthesis and analyze to understand the system itself, to understand its external context and to apply the cognitive steps to problem solve. Developing rubrics from the Systems Thinking Hierarchy (STH) model, as their assessment instrument the researchers found concept mapping to be highly effective in the students' system thinking ability especially in regard to sustainability science and engineering (Tripto, 2013).

In engineering education a quasi-experimental study was conducted to quantify the effectiveness of concept maps in learning physics in engineering degrees, using a post-test as a measuring instrument. The sample included 114 university students from the School of Industrial Engineering who were divided into two equivalent homogeneous groups of 57 students each. The researchers then conducted a crossover study with the two groups of students, with one group acting as the experimental group and another as the control group with an evspaluation instrument of 100 dichotomous items. After submitting the resulting data to a comparative statistical analysis, the study revealed a signify In cant difference in the amount of learning attained by the experiment group students as compared with the control students, with the average improvement of the experimental group was greater than 19 percentage points (Martinez, 2013).

For Novak, concept mapping was the primary tool that would make constructionism real and applicable to transform education from its previous epistemological foundation in behaviorist reductionist education. Novak's goal was to transform learning from a model which emphasized rote memory and problem solving repetition as the primary means of acquiring knowledge to the constructivist goal of meaningful learning through self-discovery and guided instruction (Novak, 2008). Figure 13, gives a multi-dimensional pattern of that new model of learning.

In Novak's multi-dimensional model of the new learning --- of which we can only give one dimension (Figure 13: Novak, 2008) ---, the science educator envisions classroom activity going from one of mainly passive rote learning (i.e. "Reception Instruction," vis a vis lectures and most textbook presentations to teaching with concept mapping for "Meaningful Learning"

Figure 13. Novak's multi-dimensional model of the new learning with concept maps (Novak, 2008)

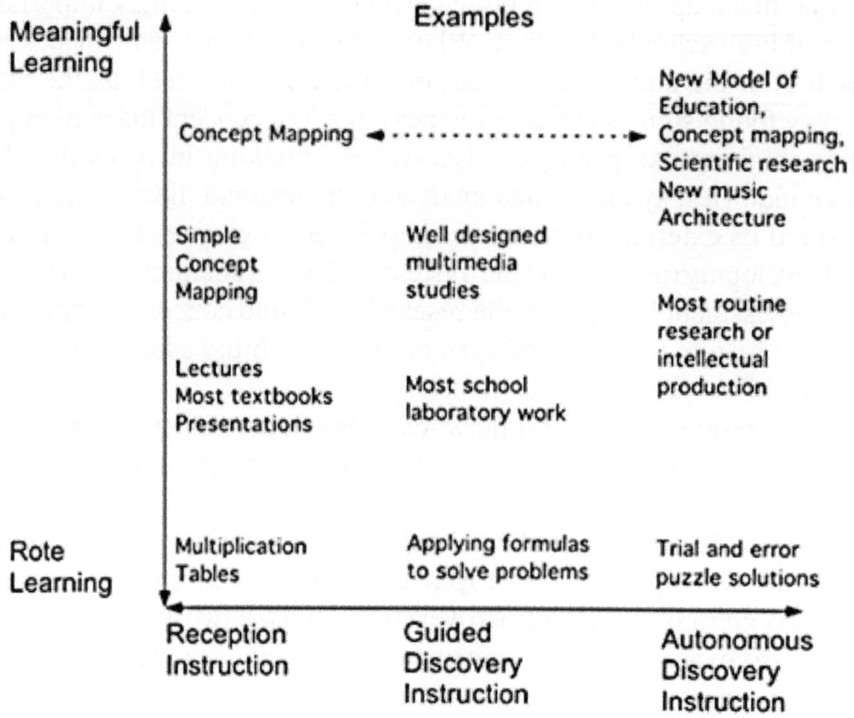

through guided discovery instruction so that students can construct their own knowledge with concept maps, which results in "Autonomous Discovery Instruction" and Learning.

For example, take a knowledge module in science concerning the seasons: instead of merely listening to lectures on the topic, reading the assigned text in a science textbook and doing the text book problems in the book as homework and then being tested on what they read and the problems they were given to solve, students would be guided to discover the scientific meaning of the seasons through advance cognitive maps on the subject and well-designed multi-media, preferably interactive studies, ending with student constructed concept maps that may appear as figure 14.

Figure 14 is a sample concept map of a student who has essentially paraphrasing a section in a text book that appears to be saying that "molecules have motion when increased by heat and that motion also determines their states which can be in the form of a solid, gas or liquid and that molecules

Figure 14. A student concept map on characteristics of molecules Herrick, 2012

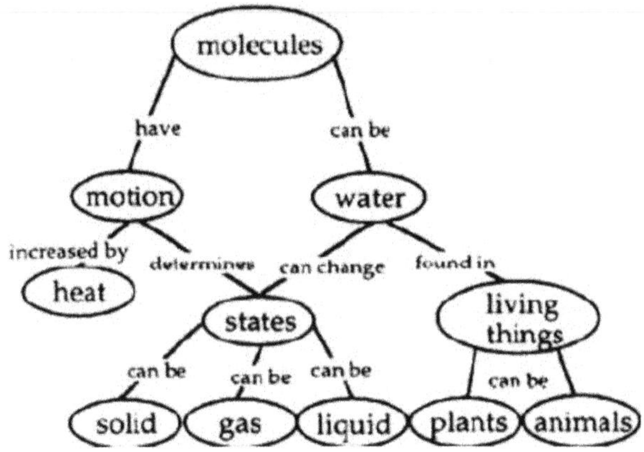

as water can change its shape aside from being found in living things such as plants and animals." From this example, which will be discussed in more detail in the next chapter, it is clear that the connection between the words as nodes (circles) represented by labelled lines, is entirely intuitive with no formal logic basis, with labels ranging from "have" to "can be" to "increased by". These intuitive and verbal aspects of concept maps will be discussed in much greater in the next chapter.

CONCLUSION

This chapter focused on the work of Albert Upton, David Ausubel and J.D. Novak as to how their ideas relate to the reinvention of verbal concept maps in the 20th century. In the case of Albert Professor Upton it is clear that his ideas of mapping out concepts camhee from a long tradition of logic which specialized to what might be called textual logic which manifested itself in two forms, linguistics (L) and semantics (S), with semantics a subspecies or subset of linguistics, or S $\subseteq$. L. However, where did Upton specifically get his idea of the four main questions that constituted the syntax of his maps? When asked this question in class, he answered, having been educated in English in the classical tradition, that it was Aristotle, and let it go at that (Upton, 1941).

Could he have been referring to Aristotle's four causes, which Ramus also tried to explicate in graphic form? This certainly is a possibility as the question, "what is this a kind of and what are the kinds (sorts) of this?" corresponds to

Aristotle's --- and therefore Ramus' --- formal cause on the essential order of things, in the sense of Aristotelian forms. The question, "what is this a part of and what are the parts of this?" is clearly from the classical tradition of material cause, in which the Ancients and the Renaissance understood as analysis according to the material structure of things. Aristotle's efficient cause, which we moderns conflate with the material cause is basically a summation of the theory of causes. In Aristotle and Ramus' day, what caused a thing to behave the way it did was its immaterial essence, or quality. Hence came Upton's question, "what is this a quality of and what are these the qualities of this?" Aristotle's final cause may correspond to "what is this a stage of and what are the states of this?' in that change is teleological, a concept positive science rejects completely. Upton's fifth method of analysis, the analog, simile or metaphor in which may be phrased in the question, "What this like and what are the likes of this?" is discussed by Aristotle apart from his four causes (Lloyd, 1979; McLuhan, 2011). Regardless as to where Professor Upton got the syntax for his textual logic, he retrieved the Aristotelian Ramus by putting that syntax in visual graphic form which later would be called concept maps

Where Professor Novak got his idea for concept mapping is much less speculative. He clearly states that – first of all – he is indebted to the historians of science, T.S. Kuhn, Stephen Toulmin, and James Byrant Conant and of course the educational psychologist, David Ausubel. J.B. Conant was a chemistry professor, education reformer, President of Harvard and Ambassador to West Germany under President Eisenhower. Given these stated influences, Novak was clearly a scholar of the revolutionary '60s.

Another important development that occurred in the 1960's was a movement from positivistic epistemologies rooted in Baconian ideas to constructivist epistemologies based on the work of Conant (1949), Kuhn (1962), and Toulmin (1972) and others. Constructivist epistemologies see knowledge creation as a human creation rooted in a social context that affect the kinds of hypotheses constructed and the research methods used. Our research group saw a close parallel between constructivist epistemological ideas and Ausubellian ideas on human learning. In time, we described an epistemology I called Human Constructivism (Novak, 2010, p 22).

Novak built his theory of concept mapping on Ausubel's constructivist theories, described in detail in this chapter, and specifically to Ausubel's theory of schema, mental constructr paradigms, that individuals, students and teachers, put together when learning through interaction with the environment. The significance and genius of J.D. Novak was he was the first to explicitly reify these schema through concept mapping. We will be discussing this highly influential educator more in the next chapter.

Furthermore, this chapter's research showed that Novak's maps were not only congruent with constructivist learning theories but also with systems thinking. As noted, in a study conducted by researchers at the Ben Gurion University in Israel on Israeli high school system, it was found that complex systems like those that make up the human body have become increasingly important as a focus of high school education. When the researchers used Novak's maps to teach both the content and structure of the systems such as the human body they found significant effectiveness in using concept maps to improve systems thinking. Having the students construct Novakian maps improved their ability to make inquiries, synthesize and analyze to understand the system itself, external context and to apply the cognitive steps to solve problems involving human systems.

REFERENCES

Abel, W. M., & Freeze, M. (2006). Evaluation of concept mapping in an associate degree nursing program. *The Journal of Nursing Education*, *45*(9). PMID:17002082

Ackermann, F., & Eden, C. (2001). Contrasting single user and networked group decision support systems for strategy making. *Group Decision and Negotiation*, *10*(1), 47–66. doi:10.1023/A:1008708912048

Ausubel, D. (1963). *The psychology of meaningful verbal learning*. New York: Grune & Stratton.

Ausubel, D., Novak, J. D., & Hanesian, H. (1978). Educational psychology: A cognitive view (2nd ed.). New York: Holt, Reinhart and Winston.

Ausubel, D. P. (1968). *Educational psychology: A cognitive view*. Holt, Rinehart & Winston.

Ausubel, D. P., & Robinson, F. G. (1969). *School learning: An introduction to educational psychology*. Holt, Rinehart & Winston.

Ausubel. (2016). Retrieved from http://www.davidausubel.org./

Bahr, G. S., & Dansereau, D. F. (2001). Bilingual knowledge maps (BiK-Maps) in second-language vocabulary learning. *Journal of Experimental Education*, *70*(1), 5–24. doi:10.1080/00220970109599496

Buzan, T. (2006). *The ultimate book of mind maps: unlock your creativity, boost your memory, change your life*. HarperCollins, UK.

Chmielewski, T. C., & Dansereau, D. F. (1998). Enhancing the recall of text: Knowledge mapping training promotes implicit transfer. *Journal of Educational Psychology*, *90*(3), 407–413. doi:10.1037/0022-0663.90.3.407

Clayton, L. H. (2006). Concept mapping: An effective, active teaching-learning method. *Nursing Education Perspectives*, *27*(4), 197–203. PMID:16921805

Conklin, J. (2003). Dialog mapping: Reflections on an industrial strength case study. In *Visualizing argumentation* (pp. 117–136). Springer London. doi:10.1007/978-1-4471-0037-9_6

Costa, A. L., & Kallick, B. (2008). *Learning and leading with habits of mind: 16 essential characteristics for success*. ASCD.

Eden, C., & Ackermann, F. (2001). Group decision and negotiation in strategy making. *Group Decision and Negotiation*, *10*(2), 119–140. doi:10.1023/A:1008710816126

Herl, H. E., ONeil, H. F. Jr, Chung, G. K. W. K., & Schacter, J. (1999). Reliability and validity of a computer-based knowledge mapping system to measure content understanding. *Computers in Human Behavior*, *15*(3), 315–333. doi:10.1016/S0747-5632(99)00026-6

Herrick, C. (2010). Challenging assumptions: Teaching, documenting, producing and negotiating health. *Journal of Geography in Higher Education*, *34*(3), 345–362. doi:10.1080/03098265.2010.498879

Hyerle, D. (1996). Thinking Maps: Seeing Is Understanding. *Educational Leadership*, *53*(4), 85–89.

Ivie, S. D. (1998). Ausubel's learning theory: An approach to teaching higher order thinking skills. *High School Journal*, *82*(1), 35–42.

Malinowski, B., Crookshank, F. G., Ogden, C. K., & Richards, I. A. (1923). *The meaning of meaning; a study of the influence of language upon thought and of the science of symbolism*. Kegan Paul, Trench, Trubner.

Martínez, G., Pérez, Á. L., Suero, M. I., & Pardo, P. J. (2013). The effectiveness of concept maps in teaching physics concepts applied to engineering education: Experimental comparison of the amount of learning achieved with and without concept maps. *Journal of Science Education and Technology*, *22*(2), 204–214. doi:10.1007/s10956-012-9386-8

Mayer, R. E. (1979). Twenty years of research on advance organizers: Assimilation theory is still the best predictor of results. *Instructional Science*, *8*(2), 133–167. doi:10.1007/BF00117008

Novak, J. D. (1990). Concept mapping: A useful tool for science education. *Journal of Research in Science Teaching*, *27*(10), 937–949. doi:10.1002/tea.3660271003

Novak, J. D. (2010). Learning, creating, and using knowledge: Concept maps as facilitative tools in schools and corporations. In Knowledge Cartography: Software tools and mapping techniques. Advanced Information and Knowledge Processing, 1. London, UK: Springer.

Novak, J. D. (2010, September). Learning, creating, and using knowledge: Concept maps as facilitative tools in schools and corporations. Invited Papers. *Journal of e-Learning and Knowledge Society, 6*(3), 21 - 30.

Novak, J. D., & Cañas, A. J. (2008). *The theory underlying concept maps and how to construct and use them. Technical Report.* Florida Institute for Human and Machine Cognition.

Novak, J. D., & Gowin, D. B. (1984). *Learning how to learn.* Cambridge University Press. doi:10.1017/CBO9781139173469

O'Neil, H. F., Jr., & Klein, D. C. D. (1997). Feasibility of machine scoring of concept maps. National Center for Research on Evaluation, Standards, and Student Testing, Office of Educational Research and Improvement (ED).

ODonnell, A. M., Dansereau, D. F., & Hall, R. H. (2002). Knowledge maps as scaffolds for cognitive processing. *Educational Psychology Review*, *14*(1), 71–86. doi:10.1023/A:1013132527007

Piaget, J., & Inhelder, B. (1969). *The psychology of the child.* Basic Books.

Raven, J. (2003). Raven progressive matrices. In Handbook of nonverbal assessment (pp. 223-237). Springer US.

Reed, C., & Rowe, G. (2001). *Araucaria: Software for puzzles in argument diagramming and XML.* Department of Applied Computing, University of Dundee Technical Report.

Rewey, K. L., Dansereau, D. F., Dees, S. M., Skaggs, L. P., & Pitre, U. (1991). Scripted cooperation and knowledge map supplements: Effects on the recall of biological and statistical information. *Journal of Experimental Education, 60*(2), 93–107. doi:10.1080/00220973.1991.10806582

Richards, I. A. (1943). *Basic English and its uses.* WW Norton & Company, Inc.

Rittel, H., & Noble, D. (1989). *Issue-based information systems for design.* Univ. Calif. Berkeley Work. Paper 492.

Shum, S. B. (2003). The roots of computer supported argument visualization. In *Visualizing argumentation* (pp. 3–24). Springer London. doi:10.1007/978-1-4471-0037-9_1

Sierhuis, M., & Shum, S. B. (2008). Human-agent knowledge cartography for e-science: NASA field trials at the Mars Desert Research Station. In *Knowledge cartography* (pp. 287–305). Springer London. doi:10.1007/978-1-84800-149-7_14

Tripto, J., Assaraf, O. B. Z., & Amit, M. (2013). January). Mapping what they know: Concept maps as an effective tool for assessing students systems thinking. *American Journal of Operations Research, 3*(01), 245–258. doi:10.4236/ajor.2013.31A022

Upton, A. (1941). *Design for thinking: A first book in semantics.* Stanford University Press.

Von Glasersfeld, E. (1995). *Radical Constructivism: A Way of Knowing and Learning.* Falmer Press, Taylor & Francis Inc. doi:10.4324/9780203454220

Vygotsky, L. S. (1978). *Mind in society: The development of higher psychological processes.* Harvard University Press.

Wertsch, J. V. (1997). *Vygotsky and the formation of the mind.* Cambridge, MA: Academic Press.

Wheeler, L. A., & Collins, S. K. (2003). The influence of concept mapping on critical thinking in baccalaureate nursing students. *Journal of Professional Nursing*, *19*(6), 339–346. doi:10.1016/S8755-7223(03)00134-0 PMID:14689390

Woolfolk, A. E., Winne, P. H., Perry, N. E., & Shapka, J. (2010). Educational Psychology (4th ed.). Toronto: Pearson Canada.

Chapter 4
Tentative Taxonomies

ABSTRACT

Because of their extensive generality, concept maps can range from the very intuitive, heuristic and free flowing on one hand to the very analytical, arcane and formulistic on the other hand with mind maps on one end of an intellectual style scale and the use of highly abstract concepts maps in computer science spatial thinking maps on the other. While the very generality of the concept of concept maps may have contributed to its increasing and extensive use in all disciplines, the range and diversity of the concept has not help in its understanding. This chapter proposes to apply two cognitive styles theories to three specific concept mapping techniques to develop tentative taxonomies which may help to increase the understanding of the nature of concept maps and how they are and can be used.

DISCUSSION

Researchers who have constructed maps aside from those mentioned in the previous chapter include Fisher (1990), Jonassen (2000), Herl (1999) and O'Neil (1999) among others (Shum 2003,Reed 2007) have used concept maps extensively in the field of computer science. Ackermann (2001) and Eden (2001) maps in which both nodes and causes can be expressed in a word, sentences or even paragraphs, .regard their concept maps as "cause maps" because the arrows that connect the nodes as indicate a causal connection. For Bahr (2001), Chmeilewski (1998), O'Donnell (2002) and Rewey (1991) concept maps are "knowledge maps" with an emphasis on scaffolding as are

DOI: 10.4018/978-1-5225-2184-6.ch004

Dialogue Maps, developed by Conklin (2003) and extended to Conversational Modelling by Martaan Sierhuis (2008). Argument maps, first proposed by J. H. Wigmore () are aids in the teaching and analysis of law with the objective of "laying out" the structure of an argument to plead a case. Parallel to Argument Maps are Horst Rittel's Issue Maps derived from an "Issue-Based Information System" (IBIS), which added a developmental function to scaffolding rules.

In addition to these versions of concept mapping, Tony Buzan has added a pictorial dimension to the technique in his Mind Maps, while David Hyerle has developed a version of concept mapping called Thinking Maps ® that can be attempted to to make concept mapping less verbal and more rule based than those previously encountered. This chapter intends to get a better handle on the ubiquity and diverseness of concept mapping by drawing up tentative taxonomies based on cognitive styles theory and practice. Specifically the chapter includes a discussion of the cognitive styles theories of Hermann Witkin and Olesya Blazhenkova and Maria Kozhevnikov (aka, B&K) on and how cognitive style theories they may be used as a basis and frames for and for constructing these taxonomies. Novak's more restricted use of concept maps, Bazun's Mind Maps and Hyerle's Thinking Maps will be a as a very limited data set to test and illustrate the tentative taxonomies which may be used for further research to gain a better understanding of this rather amorphous topic.

Buzan Maps

Tony Buzan's Mind Maps are often created around a single concept, drawn as an image in the center of a blank landscape page, to which associated representations of ideas such as pictures and words are added as concept nodes radiating from the center. Mind Maps, as made popular and taught by Tony Buzan, mind maps are usually drawn by hand and could be considered a spider maps because of its structure where branches with nodes radiate out from a central idea, theme, or target as per Figure 1 (Buzan, 2005).

Anthony "Tony" Buzan (1942 -) is an English author and educational consultant, who popularized the idea of "mental literacy" and mind mapping as a technique to improve mental performance. Buzan was born in England, went to secondary school in Vancouver, Canada, Palmers Green, where he was Head Boys' Prefect 1959-60 and completed his undergraduate studies at the University of British Columbia, where he spent a year as a graduate

Figure 1. Hand drawn mind map. Image source: http://www.mindmapart.com/health-mind-map-jane-genovese/

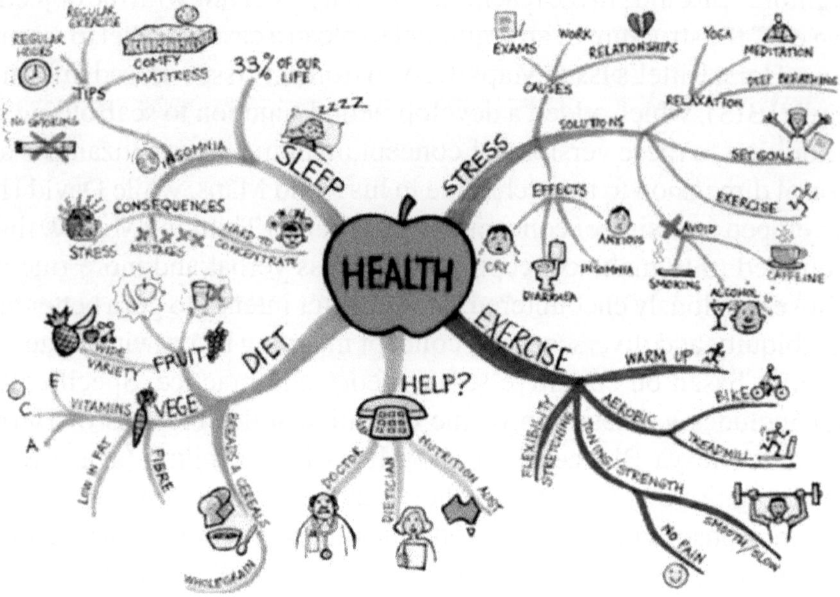

student (Mcleon,2015), and where he became very involved in Mensa. the largest and oldest high IQ society in the world and open to people who score at the 98th percentile or higher on a standardized, supervised IQ or other approved intelligence test (https://www.mensa.org/). Along with popularizing the idea of mental gymnastics to improve the brain and mind mapping he is an advocate of speed reading (Buzan, 1984, 1988, 2006). Buzan's literary, Mensa background and speed reading background leads one to conclude that his maps may have begun with a strong verbal bias and could be a reaction against that bias. Although, critics have called mind mapping pseudoscience, questioning the existence of evidence supporting the technique and the usefulness of mind mapping (Wikit, 2016), various educators have claimed empirical and experimental evidence to support its efficacy in learning across different disciplines.

For example, in an experiment in the London School of Medicine in England, researchers taught 25 students mind mapping as a study technique and compared their recall information with a control group and found that the mind mapping group, especially those who were motivated, increased their recall ability at least 10% with no adverse effects (Ferrand, p. 202).

Analysis of the data indicates that, as a strategy to improve memory for written information, the mind map technique has the potential for an important improvement in efficacy. With the mind map technique there is an estimated increase of 10% [in factual information recall] without any change in motivation, and an increase of 13% if there were such a change (p. 429).

In another experiment on mind mapping researchers found that this technique appears to increase and improve students' desire and ability to work in in teams for collaborative learning in addition to increasing their mental performance.

Encouraging is the fact that students preferred to work with mind maps in teams Mind mapping is a non-linear thinking technique which attempts to meaningfully connect two or more things/concepts that seem to be logically and rationally impossible to link (Buzan and Buzan, 1996). This rational is far away from the traditional "step process" or "linear thinking" that our students are familiar with. While working in teams for mind map construction, students have the opportunity to creativity development make the "strange familiar", to develop synergistic interaction, assemble collective knowledge and a group minded attitude (pp. 337-378).

Investigators of other experiments with Mind Maps also showed Further review of other experiments showed that the technique increased the motivation of students to master lessons and assignments related to the class or course reading material (Jones, 2012; Zampetakis,2007).

This aspect of mind mapping in regard to increasing motivation, engagement and collaborative learning, however, does not appear to be limited only to Mind Maps. Another study on a different form concept mapping researchers found similar increases in engagement and motivation along with increased knowledge acquisition. (Nesbit, 2006).

The meta-analysis found that, in comparison with activities such as reading text passages, attending lectures, and participating in class discussions, concept mapping activities are more effective for attaining knowledge retention and transfer. Concept mapping was found to benefit learners across a broad range of educational levels, subject areas, and settings. Much of this benefit may be due to greater learner engagement occasioned by concept mapping in comparison with reading and listening, rather than the properties of the concept map as an information medium (434).

Mind Maps as presented by Tony Bazun, should be contrasted with the highly technical maps presented in the previous, which in comparison appear to be crude, naïve and unsophisticated. Nevertheless, such maps have been taken seriously by scholars, professional educators alike. For example, John W. Budd (2004) a professor of management at the University of Minnesota uses mind maps to facilitate collaborate learning:

A Mind Map is a form of an outline with ideas and pictures radiating out from a central concept . . . The creation of a Mind Map in a small group is an active and collaborative learning exercise. Because a Mind Map captures a specific topic in a nonlinear fashion and incorporates graphics and colors, this exercise also can connect with learners whose style is not as well-served by traditional linear, text-based materials. Therefore, using Mind Maps as an inclass exercise is one more tool that economics instructors can add to their portfolio of active learning activities as they seek to move beyond "chalk and talk" (p 35).

The wide range of concept maps in the current environment necessitates a working taxonomy of the term, which this chapter attempts to find. The next map specimen that needs to fit into such a taxonomy is called "Novak Maps," to distinguish this kind of concept map from other such maps as Buzan's.

Novak Maps

Joseph Donald Novak, whom we have already met extensively in the previous chapter is an American educator, Professor Emeritus at the Cornell University, and Senior Research Scientist at the Florida Institute for Human & Machine Cognition. Professor Novak received a B.S. in Science and Mathematics in 1952 from the University of Minnesota, an M.S. in Science Education in 1954 from the University of Minnesota, and completed his graduate studies in Science Education & Biology at the University of Minnesota in 1958. After graduate school, Novak taught biology at Kansas State Teachers College at Emporia from 1957 to 1959, and biology and teacher education courses at Purdue University from 1959 to 1967. From 1967 to 1995, he was Professor of Education and Biological sciences at Cornell University and since 1998 he has also been Visiting Senior Scientist, at the University of West Florida, Institute for Human & Machine Cognition (IHMC,

Robert Novak's strong background and research in science education may partially explain why his maps are so popular among educators with similar backgrounds or who are using his maps for the study and teaching of science and science based disciplines such as nursing and engineering., su who concept maps unlike object maps, therefore are highly verbal in which concepts are usually enclosed in circles or boxes and relationships between concepts are indicated by a verbally labelled connecting line linking two conceptual nodes (Novak, 2006). Figure 7 is an example of a partial Novak concept map that describes the structure of his concept maps and illustrates their characteristics.

For example, Blunt and Karpickle (2014) found that Novak's concept mapping worked better among undergraduate students than the tradition use of writing for retrieving information. Funded by grants from the National Science Foundation and the United States Department of Education, Blunt and his colleague Karpicke designed and implemented an experiment with 39 Purdue undergraduate students to test the hypothesis that Novak mapping techniques for more effective for the retrieval of information from assigned text that the more traditional method of retrieval vis a vis summarizing and rephrasing the text through writing a paragraph (Karpicke, 2011; Roediger, 2006).

Learning is often viewed as a process that occurs primarily when people encode or study new material, and the best learning is thought to occur when students elaborate on what they are studying by forming meaningful connections and creating enriched knowledge structures. Retrieval, which occurs when students take tests, is viewed as an assessment of learning that occurred in prior study experiences but is not thought to create learning itself. In contrast to the latter assumption, a great deal of recent research has shown that practicing retrieval creates long-term, meaningful learning, sometimes even more learning than elaborative encoding activities . . . . Our purpose in this article was to examine the effectiveness of two different retrieval practice formats: retrieving by writing information in paragraph format (a common way to induce retrieval practice . . . and retrieving by creating what we refer to as retrieval-based concept maps (Blunt, 2014, p. 849).

After post intervention measurement, the experimenters found that

Retrieval practice is a powerful way to enhance long-term meaningful learning of educationally relevant content. The present results show that practicing retrieval, either by creating concept maps or by writing down the material in paragraph format, enhanced long-term learning more than completing the

Figure 2. Results from Blunt and karpickle's experiment showing that concept mapping as an information retrieval technique works better than traditional methods of paraphrasing and summarizing (Blunt, 2014)

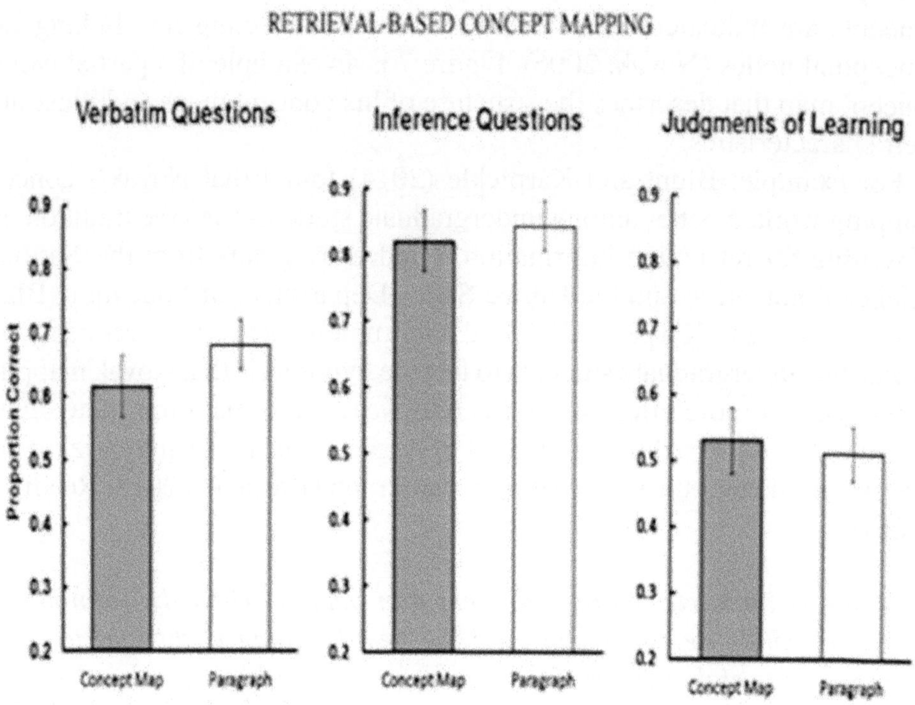

same tasks as study activities. The locus of these learning effects was in the act of retrieving knowledge, rather than the mere act of writing down. the material in paragraph or concept map format. It is important to note that the results show that concept mapping can indeed serve as an effective task when it is implemented as a retrieval-based learning activity. The key element for promoting meaningful learning was not the format of the activity; it was the requirement to engage in active retrieval practice during learning (857).

In another cases, Novak's concept maps are effective in promoting greater achievement not only across the curriculum (Yang, 2013; Wheeler, 2003), but across cultures. In an experiment on the impact of concept maps on Taiwanese 11 graders in the context of project based learning researchers found that both

Figure 3. An example of a highly verbal Novak Map used in nursing. Partial iImage source: http://www.austincc.edu/adnfac/collaborative/step3.jpg

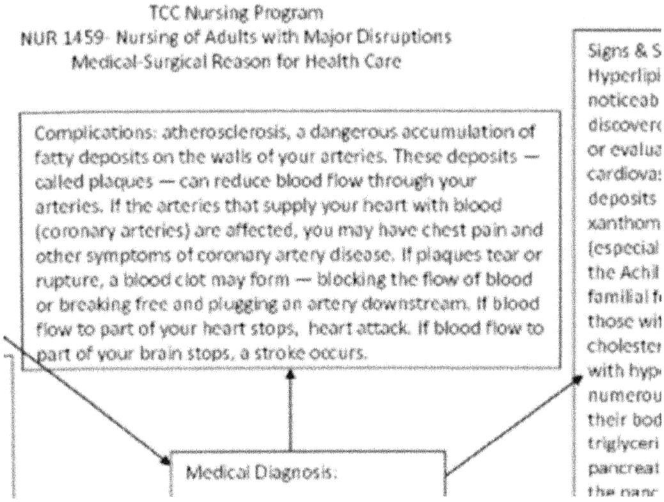

learning achievement and learning attitudes could be increased using concept mapping techniques in what are called the 5C abilities: communication, critical thinking, complex problem solving, creativity, and collaboration provided that lessons with concept mapping involved scaffolding (Wang, 2016).

One observation, however, should be noted in regard to Novak maps as distinct from as Mind Maps, Cause Maps, Argument Maps and the like, is that the authors are struck with how verbal these maps tend to be compared with other kinds of maps such as Bazun's Mind Maps. Moreover, Novak's maps emphasize the nodes as concepts rather than the arrows that connect the nodes as to how they are related or connected. Figure 36 is an example of such a highly verbal concept map that emphasizes the nodes over the relational arrows, which in this particular case are not labelled, leaving the "reader" of the map to infer the meaning of relationships. David Hyerle's Thinking Maps ®, on the other hand tend to be much less verbal, to emphasize links as relational rules that connect the conceptual nodes and, as will be argued, are more iconic in their manifestation, making them more similar to the mathematical/logical maps of Euler and Venn discussed in the previous chapters.

Hyerle Maps

Hyerle's maps, as in the case of Buzan's mind maps and Novak's concept maps have also proven equally effective in formal education settings and among different learning groups (Hyerle 2011; Marshal, 2011; Wolfe, 2011). However, these maps contrast with Buzan's maps and Novak's mainly verbal maps in that Hyerle's Thinking Maps © involve a set of abstract visual conventions designed to support core cognitive skills in the form of eight graphic structures that express basic cognitive processes such as causality, sequence, whole-part relations much like symbols in organic chemistry or in Venn diagrams (Okada, 2008; Tang, 2012).

Similar to Buzan, Hyerle originally came from a back ground that was verbally biased. After gaining his elementary and secondary teaching credentials from the Bay Area Writing Project and teaching in the Oakland and Berkeley public schools, David received his doctorate from UC Berkeley with the renowned linguist, George Lakoff (2008) on his dissertation committee, along with eminent educator, Arthur Costa (1994). Later Hyerle continued his studies as an Exchange Scholar for two years at the Harvard School of Education before becoming an international leader in the field of thinking skills development, critical reflection, and 21st century learning. Hyerle's groundbreaking book, *Visual Tools for Constructing Knowledge* (1996), focused on the comprehensive, researched based practices of how teachers facilitate creativity, analytical thinking, conceptual understanding, and "systems thinking" using his Thinking Maps. The Association for Supervision and Curriculum Development (ASCD) honored this work by distributing this book to over 120,000 educators in the United States and abroad, where he is more appreciated. As with Buzan, and to some degree, J.D. Novak, as an educator, Hyerle tends to work outside of the established ivory towers of academe and rigid confine of specific disciplines, attempting to integrate the fields of cognitive-neuroscience, Piagetian constructivist practice, and rigorous assessment into the fabric of everyday classroom teaching. Recent published writings by David include insights into assessment, knowledge cartography (knowledge creation and transfer), visual representations of data and the direct facilitation of dispositions of thinking, or *what Costa calls, habits of mind* (Hyerle, 2016).

Figure 4 shows Hyerle's Thinking Maps in their highest and most abstract form with, each of the maps operating as a symbolic structure that non-verbally indicates an operation to be performed.

Figure 4. Hyerle's Thinking Maps, which if used as non-verbal glyphs in their construction should activate the object area of the brain rather the verbal processing area. . Source: http://dft.designsforthinking.com/wp-content/uploads/2013/05/thinking_maps_logo.jpg

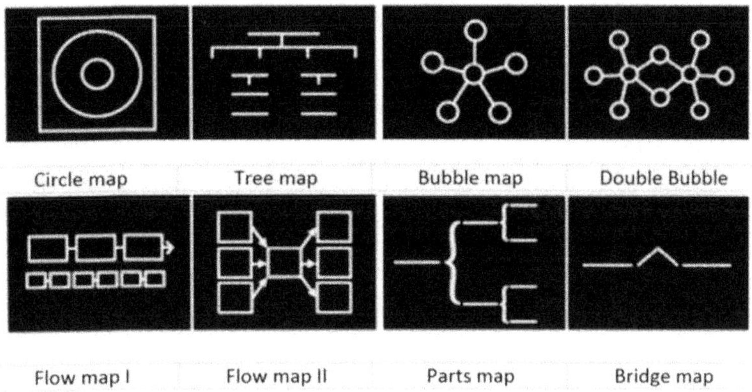

For example in the above (figure 4), when students consider or construct a circle map (upper left hand corner) they know that they are defining something in context, with that "something" being an independent variable, usually placed at the center of the map. When they see a Tree Map (upper row), the image immediately signals an ordering process involving classifying or categorization hierarchically. In constructing a bubble map (two maps in the first row upper right hand corner), students intuitively know that they are defining a variable according to its qualities or characteristics. In the same sense, the flow map and multi-flow maps (second row, lower left hand corner) indicate sequencing and/or cause and effect. The bridge map (bottom right) simulates the mental operation of making analogies for academic problem solving and real life.

The use of Thinking Maps in the classroom has proven effective in the United States, globally, in academic journals and in the field from Mississippi, Singapore to Ethiopia (Hyerle, 2011; Oklver, 2009; Long, 2011), and reasons for their success is, that unlike less structured concept maps such as many more amorphous Novak maps and Buzan maps have been summarized in figure 5.

According to the figure (5), graphic organizer type concept maps, which could include both Mind Maps and Novak Maps of the less verbal variety tend to be based on isolated tasks, use inconsistent and unrelated graphics, are often static in form, are difficult to transfer across disciplines and are usually text or teacher oriented. On the other hand, Thinking Maps are based

Figure 5. Thinking Map showing the difference between concept maps that tend to be merely graphic organizers and highly structured Thinking Maps. Image source: https://www.pinterest.com/madawn/thinking-maps/

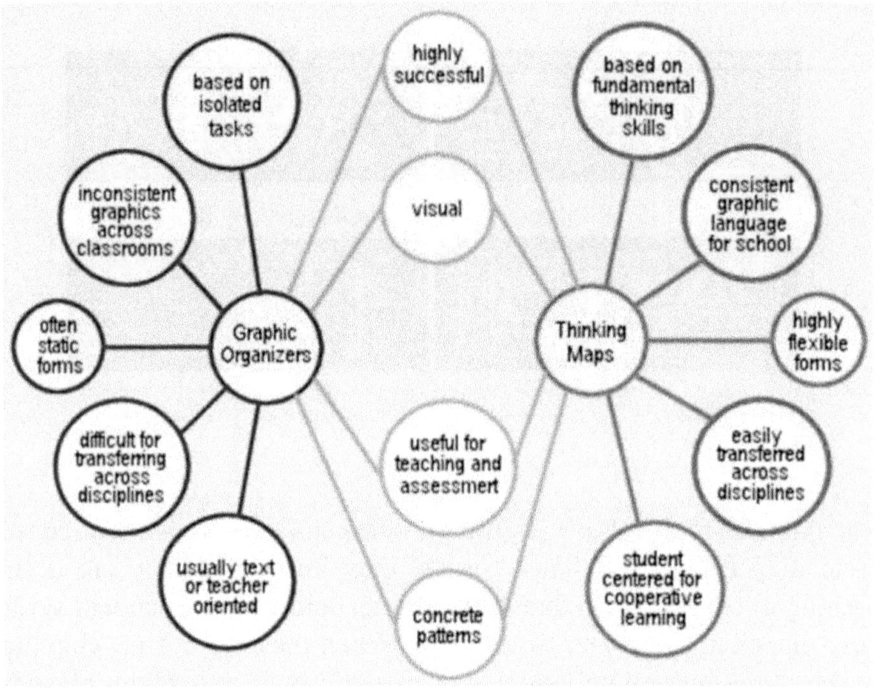

on fundamental thinking skills, uses a consistent graphic language and is easily transferred across disciplines.

In addition Hyerle and Tang (2012) have deductively proven the validity of the Thinking Maps, by reducing their structures to formal logical notation and statements to determine the reasons behind their success as teaching tools across the curriculum. After reducing all of the maps to the primitive statement $A \subseteq B$, the paper concludes that the reason for the effectiveness of Thinking Maps is because the maps themselves are graphic representations of logical processes that are not just "basic thinking skills" but of the mental operations found in set theory, cognitive semantics and perhaps even the neurological structures. The next section undertakes to classify these three types of maps, mind maps, Novak concept maps and thinking maps by associating the maps with different cognitive styles according to two neuropsychologists, O. Blazenkova and M. Kozhevnikov, after discussing other cognitive style theories for context.

Cognitive Styles Theories

Given the foregoing analysis of of three types of concept maps, Mind Maps, Novak Maps and Thinking Maps, this chapter proceeds to connect these maps with an extremely small data set (n=3) to help in the construction of two taxonomies to better understand the reason behind the effectiveness of concept maps in general.

Cognitive style, intellectual style or "thinking style," a key concept in education, is a term used in cognitive psychology to describe the way individuals think, perceive and remember or process information and perhaps different from cognitive ability in that cognitive style is malleable and therefore teachable or trainable. (Evans, 2005). Cognitive style differs from cognitive ability in that cognitive ability is measured by aptitude tests or so-called intelligence tests and can be seen to complement cognitive ability as a key concept in education, a concept that can be divided into two categories, bipolar and multi-dimensional.

Several models exist in the bi-polar category including Hudson Cary's convergent versus divergent theory and models derived from Ornstein's hemispherical lateralisation concept, popularly known as the left-brain/right-brain theory (Cary, 1991). Cognitive style nodels derived from the bilateral theory include those of Allinson-Hayes (1996) and William Taggart (1988; Robey, 1981). Perhaps better known is is Richard Riding's two-dimensional cognitive style theory along with is measuring instrument, the Cognitive Style Analysis (CSA). The CSA measures, how individuals organize, structure and process information on two orthogonal dimensions: Wholist-Analytic (W-A) and Verbal-Imagery (V-I) (Rayner, 1997). Persons described as Wholists, according to the theory, tend towards and retain a global or overall view of information while Verbalisers process and represent information in words or verbal associations, opposed to their analytical Imagers, who represent and process information in mental pictures (Kirton, 2003).

Still another dyadic cognitive styles theory is H. A. Witkin's field dependence-independence model, which identifies individual perceptive behavior by determining how well a person can distinguish specific figures from the content field in which they are set (Witkin, 1990). According to Witkin and his colleagues, individuals who exhibit field dependence tend to rely on information provided by the outer world, the field or frame of a situation on which their cognition is while field independent persons have a much narrower field (Witkin, 1975; Goodenaugh, 1976). In the context of science and engineering practice and education these two different relations to fields represent two approaches to the environment that can be characterized as reductionist versus systems; individual versus collaborative; introverted versus extroverted; atomistic versus holistic; and a preference for finding meaning in objects as parts rather than their gestalt. (Tinajero, 1997; Stansfield, 1993; Riding, 1991).

The main measurement test to determine whether a person is field independent or field dependent is the embedded figures test (EFT), in which participants are shown a simple geometric figure and then asked to find the simple figure embedded or hidden in a relatively complex figure, by circling the simple figure in the more complex one. Individual scores are determined by how many times the participant correctly identifies the embedded figure, with a higher number of correct choices indicating a greater degree of field independence.

The test is timed and its underlying learning theory is that field-independent people can better ignore the influence of the background image and find the hidden figure, while field-dependent people will find it harder to ignore the field and will be slower to locate the figure (Witkin, 1967; McKenna, 1984).

Witkin's Field model has parallels to Kolb's multi-dimensional personality inventory's bi-polar types, where Witkin's Field Independence can be associated with 1) Kolb's Reflective assimilator types such as mathematicians, chemists

Figure 6. Items from the EFT: In each item, the participant is asked to pick the figure, a, b, c, or d that contains the geometric figure x. Sources: http://www.gyanjosh.com/admin/images/upload_image/1366377815-1.bmp) and_http://www.indiabix.com/_files/images/non-verbal-reasoning/embedded-images/53.png)

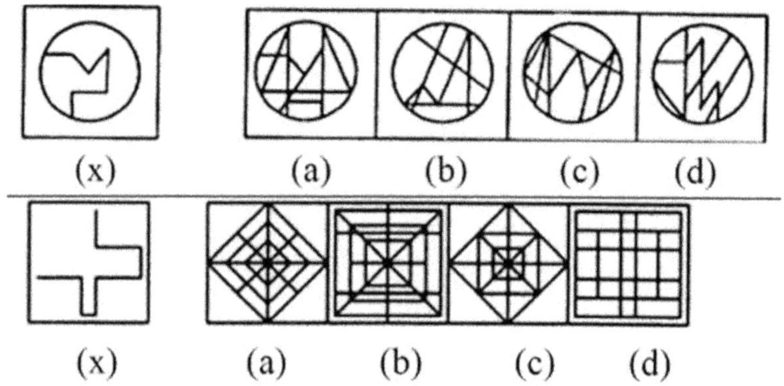

and physicists while field dependence is associated with reflective divergent types such as historians, psychologists and political scientists. (Kolb, 1981) and 2) Field dependence with Riding's Wholistic type and field independence with his Verbalizers. In the McLuhan scheme of things, Verbalizers have been conditioned by the intellectual milieu of the printing press and the Ramean textbook while Wholistic personality is the type of the present and the future as a consequence of the new electronic media technologies.

A basic understanding of the Witkin field dependent/independent theory in the context of other bi-polar cognitive styles theories allows for the construction of a tentative taxonomy based on the three types of concept maps previously discussed in detail (figure 7).

In a Witkin scheme, shown in the above figure (7) the classification of the three types of maps, Bazun, Novak and Hyerle yielded a 0, (set theory 0 = null void) for the category of field independent concept maps. Accordingly, if the Witkin classification indicates that a main characteristic of all conceptual models is that they tend to be field dependent in their construction. Good concept maps by definition are field dependent in that such maps work because they organize and process information as a function of relationships and connections which is at the core of field thinking. However, with the exception of the Hyerle maps, Buzan and Novak maps do not adequately take into consideration the importance of fields and subfields, or domains and subdomains, Hyerle Maps do, a topic that will be discussed in more detail in the next chapter. Nevertheless what is different about concept mapping as a method is that it is opposed or perhaps complementary to the standard reductionist scientific method that breaks down objects and events into smaller

Figure 7. Classification of Bazun, Novak and Hyerle according to Witkin's field dependent/independent theory

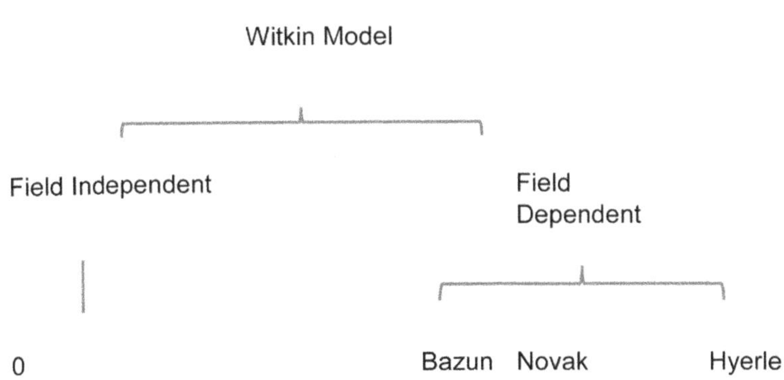

and smaller parts and arrives at conclusions or solutions one step at a time in a strictly linear progression. Concept mapping as method appears to be an alternate or complement to the traditional way of processing information by putting together instead of breaking them down and connecting them to each other in a meaningful way with an emphasis on structures. This field dependent method of analysis shifts the emphasis of finding meaning by focusing on connections, relations and wholes rather than constituent parts.

Multi-Dimensional Cognitive Styles

This chapter's analysis of Witkin's cognitive style theory shows that all three concept maps that were discussed, mind maps, Novakian maps and Hyerle's maps, operated as a field dependent cognitive mode, which when integrated with the multi-dimensional cognitive styles theory of neuropsychologists, O. Blazhenkova and M. Kozhevnikov, shows that concept maps are conductive to what the neuropsychologists call spatial thinking as opposed to more field independent modes such as verbal and numerical cognition.

The most well-known multi-dimensional cognitive types theory is the Meyers-Briggs personality indicator, the MBTI which breaks persons up into different temperament types (Meyers, 1998;.figure 8) with major categories of extroverted versus introverted (Meyers, 1998), which should be distinguished from cognitive styles theories.

Figure 9 is a Myers-Briggs chart of different personality types that has a curious resemblance of Lull's colored diagrams in the middle ages, although the wheels within wheels in the modern diagram are not meant to move.

Instead of nested circles with label parts, Figure 9 is a table of Myers-Brigg personality types. The letters can be combined to indicate different types such as ENTP would be a person with an extroverted, intuitive, thinking, perceiving characteristics and a person who is expressive, introspective, tough minded and probing in the way that person interacts with other people. On the other hand an ISFJ person has an introverted, sensing feeling and judging personality who tends to be introverted, observant, friendly and judgmental. A person can find out what type of personality she or he is by taking the Myers–Briggs Type Indicator (MBTI), an introspective self-report questionnaire designed to indicate psychological preferences in how they perceive the world and make decisions (Meyers, 2010).

While the Meyer's-Briggs typology involves personality types, another multi-dimensional theory is the cognitive theory postulated by O. Blazhenkova

Figure 8. Mind (left) and table diagram (right) of the Myers-Briggs Type Indicator. Sources (left: right: http://figures.boundless-cdn.com/11985/full/4664199330-aa06c92904-o.jpeg

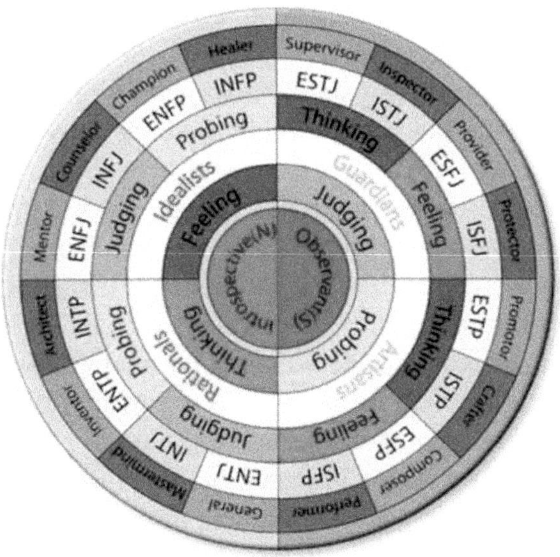

Figure 9. A table of different Myers-Briggs personality types, where letters can be combined to indicate different types. Source: http://figures.boundless-cdn.com/11985/full/4664199330-aa06c92904-o.jpeg (figure in previous page)

The Myers-Briggs ™ Type Indicator

(The Keirsey Temperament Sorter)

E Extroverted (Expressive)	S Sensing (Observant)	T Thinking (Tough-Minded)	J Judging (Scheduling)
I Introverted (Reserved)	N Intuitive (Introspective)	F Feeling (Friendly)	P Perceiving (Probing)

and M. Kozhevnikov (2010). Their theory, however is not a psychological personality theory, but a theory of cognitive styles, which also should be distinguished from cognitive aptitude theories, which claim to measure cognitive ability such as IQ tests (Flynn, 1987) and academic achievement tests such as SAT and GRE (Centra, 1971). .

Cognitive style or "thinking style," therefore is a term used in cognitive psychology to describe the way individuals think, perceive and remember information. Cognitive style differs from cognitive ability which is measured by aptitude tests or so-called intelligence tests. During its earlier stages of development cognitive styles preferences were usually binary, measuring two kinds of styles. One such cognitive style theory is that of Richard J. Riding, who divided cognitive styles into visual versus verbal (Riding, 1997; Sadler-Smith, (1999). Blazhenkova and M. Kozhevnikov departed or modified binary theories such as Riding's by formulating a tripartite multiple styles theory that broke down visualizers into persons who tend to prefer either an object or a spatial mode for processing information. It is this cognitive style theory that will be used to classify the three concept maps presented as Mind Maps, Novakian concept maps and Thinking maps.

In their new taxonomy, object visualizers prefer to construct vivid, concrete and detailed images of individual objects and rely primarily on visual-object tasks involving tactile manipulation of concrete objects. On the other hand spatial visualizers prefer to schematically represent spatial relations of objects and spatial transformations; rely primarily on visual spatial strategies and are better on spatial imagery tasks (Kozhevnikov, 2010).

B & K cognitive style theory is that it appears to be firmly grounded in brain science that neurological studies indicate that there appears to two "streams" involved in the visualization process, which Blazhenkova and Kozhenikov labelled object visualizers and spatial visualizers. . Object cognition involves processing information form a source in the visual cortex via a path called the ventral stream and the ends up in the inferior temporal lobe while spatial cognition starts at the same source in the visual cortex but takes a different path called the dorsal stream and ends in the parietal lobe. The ventral stream processing, also known as the "what pathway" is responsible for object identification and recognition, while the dorsal stream, also known as the "where pathway" is involved the spatial location and orientation of objects, giving a concrete physiological basis for the object – spatial hypothesis (Goodale, 1992). That object and spatial learning appear to be orthogonal in that individuals who score high on object style tests also tend to score low on spatial tests.

Table 1. Neurological location of different cognitive styles in different areas of the cortex. Image source: https://media.eurekalert.org/aaasnewsroom/2010/FIL_000000000865/OSV_2009.pdf

		Visual	
Systems	Verbal	Object	Spatial
Brain areas involved in processing (Cabeza & Nyberg, 2000; Gazzaniga, 2004)	Primary Auditory Cortex, Broca's area, Wernicke's area, Supramarginal gyrus, Angular gyrus	The ventral stream runs through V1, V2, V4 and to the inferior temporal lobe	The dorsal stream runs through V1, V2, V3, V5 (also known as MT area) and to the inferior parietal lobe
Functions (Farah et al., 1988; Paivio, 1991)	Representation of verbal information, comprehension and production of spoken and written language	Representation of individual objects and scenes in colour, detail and shape; object recognition	Representation of object locations, motion, spatial relationships, manipulations and transformations
Units and organization (Paivio, 1991; Richardson, 1977)	Discrete linguistic units	Integer visual scenes and discrete objects	
	Symbolic representations systematically organized semantically and grammatically	Pictorial, analogous, representations of objects and scenes	Schematic representations of objects and patterns, and their transformations
Processing (Kozhevnikov et al., 2005; Paivio, 1991)	Sequential	Holistic	Sequential
	Units have to be integrated in order to be comprehended	Units have to be identified and separated in the continuous visual field	
	Phonetically processed before assessing semantic representation	Meaningfully identified before it can be verbally named	Not necessarily identified and verbally named

Moreover, verbal style thinkers prefer to represent and process information through words and texts, using discrete linguistic units, ordered according to sequential semantic and grammatical structures. Visual object style thinkers, on the other and prefer to process information through the representation of individual objects and scenes in color, detail and shape in a holistic frame. Witkin, would call verbal thinkers field independent while object style thinkers would comprehend information as field dependent. Spatial style thinkers are also field dependent and prefer the representation of objects through their location, motion, spatial relationships which can be manipulated and transformed. The spatial cognitive style differs from the object style in that spatial cognition is much more abstract, and similar to the verbal style, involves the process of information sequentially.

Both object visualizers and spatial thinkers are opposed to verbal thinkers who process information in the left side of the brain and who appear to have a low preference for object learning but a much less aversion for spatial cognition, as illustrated in figure 10, resulting in a tripartite multiple style theory that could be called an "object-spatial imagery and verbal theory" (Kozhenivov, 2005)

To conduct further research on their model that connected cognitive style preferences to areas in the brain, they developed an Object-Spatial Imagery Verbal Questionnaire (OSIVQ) to tie their styles to different professions and found that persons in different professions had different style preferences. For example, artists preferred object thinking, social scientists, verbal thinking and scientists and engineers, spatial cognition, findings that have been verified by other researchers (Kozhevnikov, 2014). Other researchers in education (Casakin and Gigi, 2014) found that successful applicants to architecture schools preferred a visual cognitive style (spatial) over the semantic (verbal) while Kawahara and Matsuoka found that different college majors have different style preferences where engineers scored high spatially but low in object thinking while art students scored the reverse with high object and low spatial scores. Medical students were the only subjects surveyed who had positive scores in both object and spatial cognition.

Figure 10. Object-Spatial-Verbal Cognitve Style Model (Kozhevnikov, 2005) with a tentative categorization of three types of concept maps

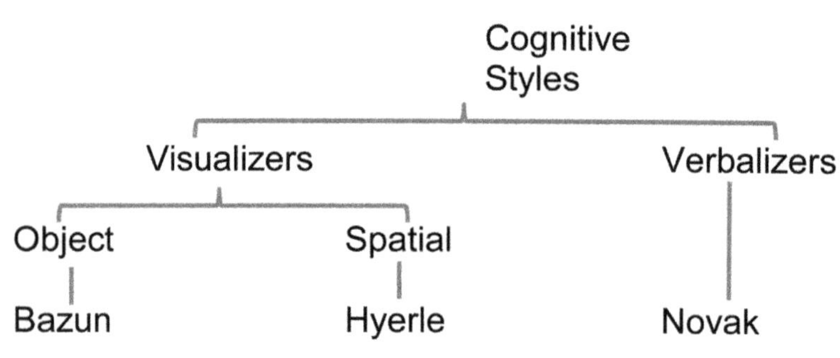

Unfortunately, the Japanese study did not measure the verbal cognitive skill set of the students and had they done so the authors believe that the medical students would have scored positively on all three of B & K's object, spatial and verbal styles, giving us an ideal profile that can be applied to the holistic systems STEM student (Kawahara, 2013). From this research, we have sufficient data to relate Blazhenkova's and Kozhevnikov's cognitive style theory to our analysis of three types of cognitive maps.

Object, Spatial and Verbal Maps

Framed by the B&K theory of cognitive style Bazun's maps can be classified as, "object maps", a kind of map that aligns with the object style of cognition that would be preferred by object learning persons. These are individuals who visualize problems in very vivid, concrete, colorful and often dynamic forms, is associated with visual artists and designers and characterized by photographic memory, implicit attention to details and emotional and aesthetic information processing (Kozhevnikov, 2005). Buzan maps often appear similar to artists' drawings and paintings in that they are strongly visual, colorful and free flowing although mind maps are often created around a single concept, drawn as an image in the center of a blank landscape page, to which associated representations of ideas such as pictures and words are added. Major ideas are connected directly to the central concept, and other ideas branch out from those and can be presented in any desired multi-media form, including video and sound liked to object nodes. .,

Although concept maps in general, whether object, verbal or spatial may increase the motivation for students to study, enhance their engagement to learn, and encourage them to work in teams, for mind mapping in particular, it is hypothesized that the concrete operational style of drawing images to organize and process information activates the same area of the brain that is activated when visual artists draw, paint or film video (Kozhevnikov, 2005).

According to Blazhenkova and Kozhevnikov, the verbal style of thinking activates the Broca's and Wernicke's area of the brain in the left hemisphere; processes information through discrete linguistic units sequentially; tends to promote linear and reductionist thinking by breaking information into isolated parts; and symbolic representations are systematically organized according to semantic and grammatical structures. Robert Novak's concept maps unlike Bazun's object maps and Hyerle's Thinking Maps are highly verbal in which concepts, often sentences and even paragraphs, are enclosed in circles or boxes and relationships between the concepts connected by labelled arrows hierarchically connecting conceptual nodes (Novak, 2006). As the case with mind mapping, strong evidence exists to show that Novak's concept maps are effective in promoting greater achievement across the curriculum (Yang, 2013; Wheeler, 2003), increasing ability in what has been called the three C's communication, critical thinking, complex problem solving, creativity, and collaboration. (Wang, 2016).

Spatial thinkers, Blazhenkova and Kozhevnikov argue, prefer to schematically represent spatial relations of objects and their transformations in time and space; rely primarily on visual-spatial strategies in problem solving, are less dependent and biased by verbal analysis, think in terms of highly abstract structures are better on spatial imagery tasks (Blazhenkova, 2009). Cognitive style researchers and educators have found that engineers, scientists architects and technicians in professions involving highly abstract abilities to mentally manipulate objects from a stationary and relative points of view excel in this kind of thinking.

Spatial maps as defined by what can be called spatial thinking of the three concept maps analyzed best fits David Hyerle's Thinking Maps ® when used by senior high school and undergraduate students (Hyerle, 2014). These maps, when used for higher order thinking, are a more rigorous version of concept mapping that limit the number of maps that can be used to eight, with each specific map, having a specific function to indicate a logical relationship presented as a specific structure. Hyerle's maps unlike the object and verbal maps approach the ideal of a proto-language and has many similarities to mathematical languages found in formal logic as manifested in diagram form such as Euler and Venn diagrams as shown in figures 11 and 12. (Tang, 2012).

CONCLUSION

Figure 14 is a tentative concept map taxonomy that results from: 1) framing the map with cognitive style theory; 2) connecting Hyerle's Thinking Maps with visual spatial Euler and Venn diagrams and 3) combining selected maps from Chapters One and Chapter Two and including several concept maps previously mentioned, namely, J.H. Wigmore's Argument Maps and Horst Rittel's Issue Maps. Before displaying the map, however, let us make a preliminary division using a simple Venn Diagram of for sets: General or meta, specific, verbal and visual, recalling that according to Blazhenkova and Kozhevnikov, verbal maps are distinguished from spatial maps, of which

Figure 11. Euler/Venn diagram with intersecting sets

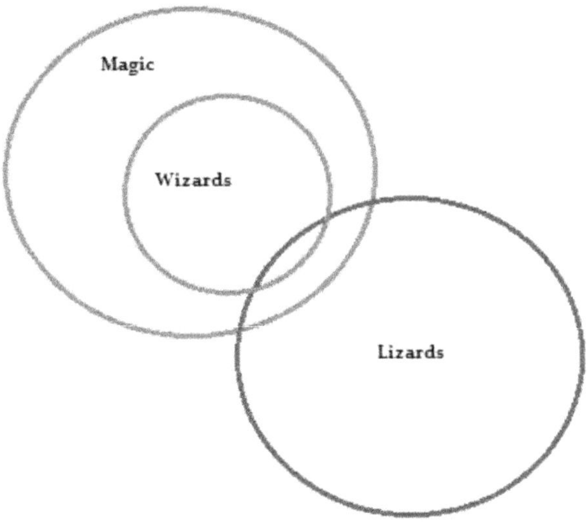

Figure 12. Venn intersecting sets expressed as Thinking Map

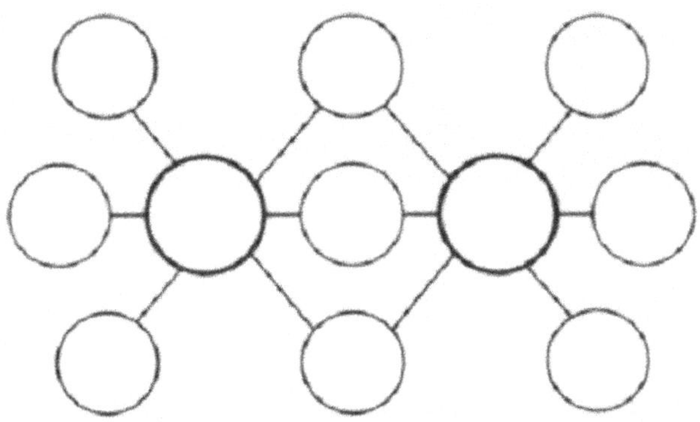

there are two: object and spatial (figure 13). In that figure meta (m) intersects verbal (v) intersect spatial (s1) intersects specific (s2) etc. Therefore the maps analyzed in this chapter can be characterized as follows: Mind Maps = b = meta ∩ visual = meta visual object maps; Novak Maps = a = metal ∩ verbal = meta verbal maps; Thinking maps = b = meta visual spatial maps. Venn/ Euler/Carroll diagrams discussed in the previous chapter could be classified as meta specific visual concept maps in that they are used as general maps for a specific domain for logical operations.

The same maps can also be classified as in Figure 14.

Figure 13. Venn diagram where meta (m) intersects verbal (v1) intersect visual (v2) and specific (s)

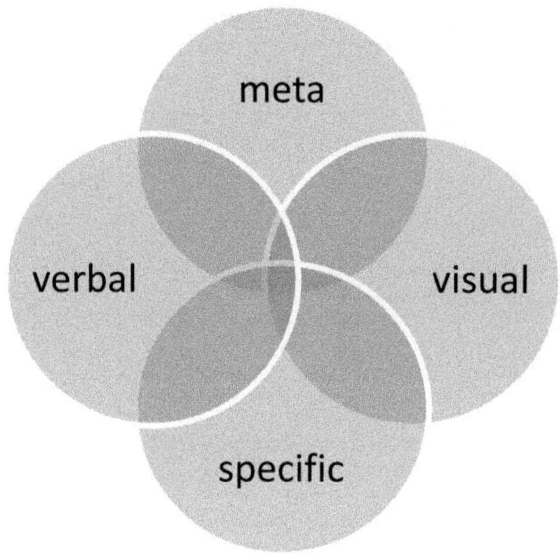

Figure 14. A taxonomy of concept maps

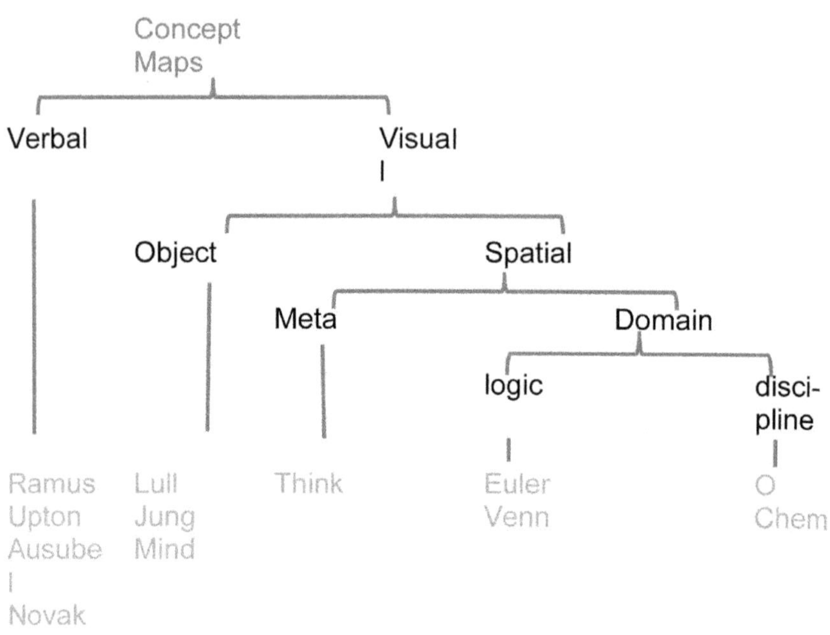

Concept maps therefore can be verbal or visual, with verbal maps beginning with Ramus and retrieved by Upton, Ausubel and Novak. Meanwhile, Lull's, Jung's and Mind Maps can are visual object maps with visual spatial maps the parent of logical maps, as represented by Euler and Venn. The two outliers in this scheme of things are thinking maps spatial as meta-maps and symbols found in organic chemistry, which a maps restricted to specific disciplines. It will be argued in the next chapter, both Thinking Maps and the symbols found in modern sciences, are the maps of the future.

REFERENCES

Abel, W. M., & Freeze, M. (2006). Evaluation of concept mapping in an associate degree nursing program. *The Journal of Nursing Education*, *45*(9). PMID:17002082

Affley, P. (2012). Putting thinking at the heart of learning.... *Primary Teacher Update*, *1*(15), 19–20. doi:10.12968/prtu.2012.1.15.19

Al-Jarf, R. (2009). Enhancing freshman students' writing skills with a mind-mapping software. In *Proceedings of eLearning and Software for Education (eLSE)* (No. 1, pp. 375-382). Universitatea Nationala de Aparare Carol I.

Blazhenkova, O., & Kozhevnikov, M. (2009). The new object-spatial-verbal cognitive style model: Theory and measurement. *Applied Cognitive Psychology*, *23*(5), 638–663. doi:10.1002/acp.1473

Blunt, J. R., & Karpicke, J. D. (2014). Learning with retrieval-based concept mapping. *Journal of Educational Psychology*, *106*(3), 849–858. doi:10.1037/a0035934

Buzan, T. (1984). *Make the most of your mind*. Simon and Schuster.

Buzan, T. (1988). *Speed reading*. David & Charles.

Buzan, T. (2005). *The ultimate book of mind maps*. Wellingborough: Thorsons.

Buzan, T. (2006). *The ultimate book of mind maps: unlock your creativity, boost your memory, change your life*. HarperCollins, UK.

Buzan. (1996). *How to use radiant thinking to maximize your brain's untapped potential*. Plume Books.

Carnot, M. J., Feltovich, P., Hoffman, R. R., Feltovich, J., & Novak, J. D. (2003). *A summary of literature pertaining to the use of concept mapping techniques and technologies for education and performance support*. Pensacola, FL: Academic Press.

Casakin, H., & Gigi, A. (2016). Cognitive styles in admission procedures for assessing candidates of architecture. *Assessment & Evaluation in Higher Education*, *41*(2), 167–182. doi:10.1080/02602938.2014.992298

Chung, G. K. W. K., ONeil, H. F. Jr, & Herl, H. E. (1999). The use of computer-based collaborative knowledge mapping to measure team processes and team outcomes. *Computers in Human Behavior*, *15*(3), 463–493. doi:10.1016/S0747-5632(99)00032-1

Clayton, L. H. (2006). Concept mapping: An effective, active teaching-learning method. *Nursing Education Perspectives*, *27*(4), 197–203. PMID:16921805

Conant, J. B. (1947). *On Understanding Science: An Historical Approach*. New Haven, CT: Yale University Press.

Costa, A. L., & Garmston, R. J. (1994). *Cognitive coaching: A foundation for renaissance schools. Christopher-Gordon Publishers, Inc.*

Davies, M. (2011). Concept mapping, mind mapping and argument mapping: What are the differences and do they matter? *Higher Education*, *62*(3), 279–301. doi:10.1007/s10734-010-9387-6

Eppler, M. J. (2006). A comparison between concept maps, mind maps, conceptual diagrams, and visual metaphors as complementary tools for knowledge construction and sharing. *Information Visualization, 5*(3), 202-210.

Evans, C. (2015). L.-F. Zhang: The malleability of intellectual styles. *Higher Education*, *69*(1), 169–172. doi:10.1007/s10734-014-9768-3

Farrand, P., Hussain, F., & Hennessy, E. (2002). The efficacy of themind mapstudy technique. *Medical Education*, *36*(5), 426–431. doi:10.1046/j.1365-2923.2002.01205.x PMID:12028392

Fisher, K. M. (1990). Semantic networking: The new kid on the block. *Journal of Research in Science Teaching*, *27*(10), 1001–1018. doi:10.1002/tea.3660271008

Fisher, K. M. (2005). SemNet software as an assessment tool. In *Assessing science understanding: A human constructivist view* (pp. 197–221). Academic Press. doi:10.1016/B978-012498365-6/50011-1

Flynn, J. R. (1987). Massive IQ gains in 14 nations: What IQ tests really measure. *Psychological Bulletin, 101*(2), 171–191. doi:10.1037/0033-2909.101.2.171

Goodale, M. A., & Milner, A. D. (1992). Separate visual pathways for perception and action. *Trends in Neurosciences, 15*(1), 20–25. doi:10.1016/0166-2236(92)90344-8 PMID:1374953

Hung, J. L. (2012). Trends of e-learning research from 2000 to 2008: Use of text mining and bibliometrics. *British Journal of Educational Technology, 43*(1), 5–16. doi:10.1111/j.1467-8535.2010.01144.x

Hwang, G. J., Yang, L. H., & Wang, S. Y. (2013). A concept map-embedded educational computer game for improving students learning performance in natural science courses. *Computers & Education, 69*, 121–130. doi:10.1016/j.compedu.2013.07.008

Hyerle, D. (2014). Thinking Maps®: A Visual Language for Learning. In A. Okada, S. B. Shum, & T. Sherborne (Eds.), *Knowledge Cartography* (pp. 73–88). London: Springer. doi:10.1007/978-1-4471-6470-8_4

Hyerle, D. (2016). *David Hyerle, Design for Thinking*. Available at: http://dft.designsforthinking.com/?page_id=199

Hyerle, D. N., & Alper, L. (2011). *Student successes with thinking maps®: school-based research, results, and models for achievement using visual tools*. Corwin Press.

Hyerle, D. N., & Alper, L. (Eds.). (2011). *Student successes with thinking maps®: school-based research, results, and models for achievement using visual tools*. Corwin Press.

IHMC. (2016). *Joseph Novak*. Institute for Human and Machine Cognition. Available at https://www.ihmc.us/groups/jnovak/

Jablokow, K. W., DeFranco, J. F., Richmond, S. S., Piovoso, M. J., & Bilén, S. G. (2015). Cognitive style and concept mapping performance. *The Journal of Engineering Education, 104*(3), 303–325.

Jones, B. D., Ruff, C., & Sunder, J. D. (2012). The effects of mind mapping activities on students motivation. *International Journal for the Scholarship of Teaching and Learning*, *6*(1), 1–21. doi:10.20429/ijsotl.2012.060105

Kawahara, M., & Matsuoka, K. (2013). Object-Spatial Imagery Types of Japanese College Students. *Psychology (Savannah, Ga.)*, *4*(03), 165–168.

Kozhevnikov, M. (2007). Cognitive styles in the context of modern psychology: Toward an integrated framework of cognitive style. *Psychological Bulletin*, *133*(3), 464–481. doi:10.1037/0033-2909.133.3.464 PMID:17469987

Kozhevnikov, M., Blazhenkova, O., & Becker, M. (2010). Trade-off in object versus spatial visualization abilities: Restriction in the development of visual-processing resources. *Psychonomic Bulletin & Review*, *17*(1), 29–35. doi:10.3758/PBR.17.1.29 PMID:20081157

Kozhevnikov, M., Blazhenkova, O., & Becker, M. (2010). Trade-off in object versus spatial visualization abilities: Restriction in the development of visual-processing resources. *Psychonomic Bulletin & Review*, *17*(1), 29–35. doi:10.3758/PBR.17.1.29 PMID:20081157

Kozhevnikov, M., Evans, C., & Kosslyn, S. M. (2014). Cognitive style as environmentally sensitive individual differences in cognition: A modern synthesis and applications in education, business, and management. *Psychological Science in the Public Interest*, *15*(1), 3–33. doi:10.1177/1529100614525555 PMID:26171827

Kozhevnikov, M., Kosslyn, S., & Shephard, J. (2005). Spatial versus object visualizers: A new characterization of visual cognitive style. *Memory & Cognition*, *33*(4), 710–726. doi:10.3758/BF03195337 PMID:16248335

Kuhn, T. S. (2012). *The structure of scientific revolutions*. University of Chicago press. doi:10.7208/chicago/9780226458144.001.0001

Lakoff, G., & Johnson, M. (2008). *Metaphors we live by*. University of Chicago press.

Lambiotte, J. G., Skaggs, L. P., & Dansereau, D. F. (1993, November). knowledge maps and cooperative review strategies. *Applied Cognitive Psychology*, *7*(6), 483–497. doi:10.1002/acp.2350070604

Latha, S., Raj, M. C., & Sukumaran, R. (2016). Use of visualization techniques for active learning: Engagement in environmental science engineering courses. International Journal of Social, Behavioral, Educational, Economic. *Business and Industrial Engineering*, *10*(4), 1276–1280.

Lloyd, G. E. R. (1979). *Magic, reason and experience*. Cambridge, UK: Cambridge University Press.

Long, D., & Carlson, D. (2011). Mind the map: How thinking maps affect student achievement. *Networks*, *13*(2), 1–7.

Malinowski, B., Crookshank, F. G., Ogden, C. K., & Richards, I. A. (1923). *The meaning of meaning; a study of the influence of language upon thought and of the science of symbolism*. Kegan Paul, Trench, Trubner.

Martínez, G., Pérez, Á. L., Suero, M. I., & Pardo, P. J. (2013). The effectiveness of concept maps in teaching physics concepts applied to engineering education: Experimental comparison of the amount of learning achieved with and without concept maps. *Journal of Science Education and Technology*, *22*(2), 204–214. doi:10.1007/s10956-012-9386-8

Mashal, N., & Kasirer, A. (2011). Thinking maps enhance metaphoric competence in children with autism and learning disabilities. *Research in Developmental Disabilities*, *32*(6), 2045–2054. doi:10.1016/j.ridd.2011.08.012 PMID:21985987

Mashal, N., & Kasirer, A. (2011). Thinking maps enhance metaphoric competence in children with autism and learning disabilities. *Research in Developmental Disabilities*, *32*(6), 2045–2054. doi:10.1016/j.ridd.2011.08.012 PMID:21985987

McLeod, B (2015, October 19). We were here first. *The Peak*, p. F1.

McLuhan, M., & McLuhan, E. (2011). *Media and formal cause*. NeoPoiesis Press.

Myers, I., & Myers, P. (2010). *Gifts differing: Understanding personality type*. Nicholas Brealey Publishing.

Nesbit, J. C., & Adesope, O. O. (2006). Learning with concept and knowledge maps: A meta-analysis. *Review of Educational Research*, *76*(3), 413–448. doi:10.3102/00346543076003413

Okada, A., Shum, B. S., & Sherborne, T. (Eds.). (2008). Knowledge cartography: Software tools and mapping techniques. Springer.

Oliver, K. (2009). An investigation of concept mapping to improve the reading comprehension of science texts. *Journal of Science Education and Technology*, *18*(5), 402–414. doi:10.1007/s10956-009-9157-3

Piaget, J. (1952). The origins of intelligence in children: Vol. 8. *No. 5* (pp. 18–36). New York: International Universities Press.

Prawat, R. S., & Floden, R. E. (1994). Philosophical perspectives on constructivist views of learning. *Educational Psychologist*, *29*(1), 37–48. doi:10.1207/s15326985ep2901_4

Raven, J. (2003). Raven progressive matrices. In Handbook of nonverbal assessment (pp. 223-237). Springer US.

Rayner, S., & Riding, R. (1997). Towards a categorisation of cognitive styles and learning styles. *Educational Psychology*, *17*(1-2), 5–27. doi:10.1080/0144341970170101

Richards, I. A. (1943). *Basic English and its uses*. WW Norton & Company.

Riding, R. J., & Sadler-Smith, E. (1997). Cognitive style and learning strategies: Some implications for training design. *International Journal of Training and Development*, *1*(3), 199–208. doi:10.1111/1468-2419.00020

Robey, D., & Taggart, W. (1981). Measuring managers' minds: The assessment of style in human information processing. *Academy of Management Review*, *6*(3), 375–383.

Robey, D., & Taggart, W. (1981). Measuring managers' minds: The assessment of style in human information processing. *Academy of Management Review*, *6*(3), 375–383.

Rowe, G., & Reed, C. (2006). Translating Wigmore Diagrams. *Frontiers in Artificial Intelligence and Applications*, *144*, 171–241.

Sadler-Smith, E., & Riding, R. (1999). Cognitive style and instructional preferences. *Instructional Science*, *27*(5), 355–371. doi:10.1007/BF00892031

Sierhuis, M., & Shum, S. B. (2008). Human-agent knowledge cartography for e-science: NASA field trials at the Mars Desert Research Station. In *Knowledge Cartography* (pp. 287–305). Springer London. doi:10.1007/978-1-84800-149-7_14

Suthers, D., Connelly, J., Lesgold, A., Paolucci, M., Toth, E., Toth, J., & Weiner, A. (2001). Representational and advisory guidance for students learning scientific inquiry. *Smart machines in education: The coming revolution in educational technology*, 7-35.

Tang, M., David, H., Byrne, R., & Tran, J. (2012). A mathematical analysis of semantic maps, with theoretical and applied implications for blended learning software. *International Journal on E-Learning*, *11*(1), 95–104.

Toulmin, S. (1972). Human understanding: Vol 1. The collective use and evolution of concepts. Princeton University Press.

Tripto, J., Assaraf, O. B. Z., & Amit, M. (2013). Mapping what they know: Concept maps as an effective tool for assessing students systems thinking. *American Journal of Operations Research*, *3*(01), 245–258. doi:10.4236/ajor.2013.31A022

Uttal, D. H., Meadow, N. G., Tipton, E., Hand, L. L., Alden, A. R., Warren, C., & Newcombe, N. S. (2013). The malleability of spatial skills: A meta-analysis of training studies. *Psychological Bulletin*, *139*(2), 352–402. doi:10.1037/a0028446 PMID:22663761

Wang, H. Y., Huang, I., & Hwang, G. J. (2016). Effects of a Question Prompt-based Concept Mapping Approach on Students' Learning Achievements, Attitudes and 5C Competences in Project-based Computer Course Activities. *Journal of Educational Technology & Society*, *19*(3), 351–365.

Watson, M. K., Pelkey, J., Noyes, C. R., & Rodgers, M. O. (2016). Assessing Conceptual Knowledge Using Three Concept Map Scoring Methods. *The Journal of Engineering Education*, *105*(1), 118–146. doi:10.1002/jee.20111

Wheeler, L. A., & Collins, S. K. (2003). The influence of concept mapping on critical thinking in baccalaureate nursing students. *Journal of Professional Nursing*, *19*(6), 339–346. doi:10.1016/S8755-7223(03)00134-0 PMID:14689390

Wolfe, P. (2011). Student Successes with Thinking Maps. In D. N. Hyerle & L. Alper (Eds.), *School-based research, results, and models for achievement using visual tools* (pp. 197–221). Corwin Press.

Yang, C. C., Hwang, G. J., Hung, C. M., & Tseng, S. S. (2013). An evaluation of the learning effectiveness of concept map-based science book reading via mobile devices. *Journal of Educational Technology & Society*, *16*(3), 167–178.

Zampetakis, L. A., Tsironis, L., & Moustakis, V. (2007). Creativity development in engineering education: The case of mind mapping. *Journal of Management Development*, *26*(4), 370–380. doi:10.1108/02621710710740110

Section 3

Concept Mapping for STEM Disciplines

One possible explanation for the popularity of concept mapping is perhaps educators are anxiously looking for alternate tools, interventions, concepts or paradigms to halt or reverse what they consider to be the attenuation of academic achievement in the United States. The failure of tradition secondary education to realize the goal of academic achievement to prepare students for college is well documented by scores on standardized tests such as the Standard Achievement Test (SAT), widely used for college admissions, which have been declining since the 1960s. Clearly an urgent need exists for new intervention education materials, approaches and strategies to improve the academic skills of students in the STEM disciplines.

Chapter 5
Visual Logic Maps (vLms)

ABSTRACT

This concluding chapter describes and analyses a concept map based, Visual Logic Maps (vLms). Essentially the vLms maps differ from the Thinking Maps discussed in the last chapter in that the Visual Logic Maps shifts its emphasis from map structure to map glyphs. In Thinking Maps it is the structure of the eight specific maps that determine how information is going to be organized and mentally process. The Visual Logic Maps reduces each of the maps to seven specific glyphs that operate as constant logical operators. In the conclusion of the chapter and the book it is argued that both the Thinking Maps and Visual Logic Maps are essentially non-verbal spatial maps that find the cognitive origins in the logical/mathematical diagrams of Venn and Euler. Unlike Venn an Euler Diagrams, however, Visual Logic and Thinking Maps are not domain specific nor mathematical in the numeric sense.

DISCUSSION

As indicated in the previous chapter, the concept of "concept maps" has become quite popular in education, largely because J. D. Novak, a constructivist educator (Piaget, 1969) scholar who developed Ausubel's idea of schemas and their construction in a research program when he and his colleagues at Cornell tried to trace and understand changes in children's knowledge of science in a less linear and more holistic way. During the course of this study Novak came up with the idea of representing children's knowledge in diagrams where concepts are linked together by arrows or lines that indicated

DOI: 10.4018/978-1-5225-2184-6.ch005

that a logical relationship existed between connected concepts (Novak, 2008). The phylogeny of Novak's maps are generally from top down and by following one branch of the map that tries to define concept mapping in figure n of the previous chapter. Through deductive reasoning a student learns that "concept maps represent [the] organized knowledge needed to answer focused questions that are context dependent." Following another branch, "concept maps represent organized knowledge comprised of concepts [and] propositions [that] are . . ." etc., compressing a large amount of meaningful information in a small amount of space.

From the previous chapters it was also found that concept maps can range from the very intuitive, heuristic and free flowing on one hand to the very analytical, arcane and formulistic on the other with mind maps on one end of an intellectual style scale and computer ontology on the other. On the highly analytical end computer science concept maps depict complex systems in visual diagrams that define and logically sort out steps, types, properties, and interrelationships of concepts needed to solve a computational problem (Guber, 1993; Katifori, 2007; Story, 2013) to the very concrete, colorful and visual Bazun Maps as shown in Figure 3 (left).

In addition, the previous section on cognitive styles showed that O. Blazhenkova and M. Kozhevnikov's theory can serve as a frame for guidance to understand concept maps from a neuropsychological perspective. That frame led to a taxonomy of concept maps associating Buzan's Mind Maps with the object style of learning, Novak Maps with verbal learning and, to a certain, degree, Hyerle's Thinking Maps with spatial learning.

For the past three or four years, researchers at the University of Colorado Denver's College of Engineering and Applied Science have been investigating concept maps as a philosophy and technique to improve learning among science and engineering students. In pursuing that goal, the investigators conducted their own experiments with concept mapping using their own concept maps called Visual Logic Maps (vLms).

These experiments at the University of Colorado Denver (UCD) on the the use of Visual Logic Maps show that they increase concept comprehension and content knowledge in STEM disciplines. In an undergraduate college general chemistry course students enrolled in the experimental classroom (n = 60) were asked to develop their own visual logic concept maps to better understand the content of readings in their chemistry textbook. Students who were enrolled in the control classroom (n = 58) studied the same learning content, did the same homework assignments and had the same course instructor, as students in experimental section. Students in the control classroom, however, were not required to learn about and construct concept maps as part of their course activities. In the chemistry class, students in the experimental class (concept mapping class) made significant learning gains in terms of both their qualitative (conceptual) and quantitative (traditional) understandings of general chemistry content (Tang, 2013). The study involved collecting quantitative evidence to support the hypothesis that greater net-learning gains are associated with student use of cognitive maps. The preliminary study focused on the measurement of content knowledge gains. Results from the preliminary study in the chemistry classes can be found in Figures 1 and 2.

Figure 1. Comparison of net-learning gains by content area and content type I

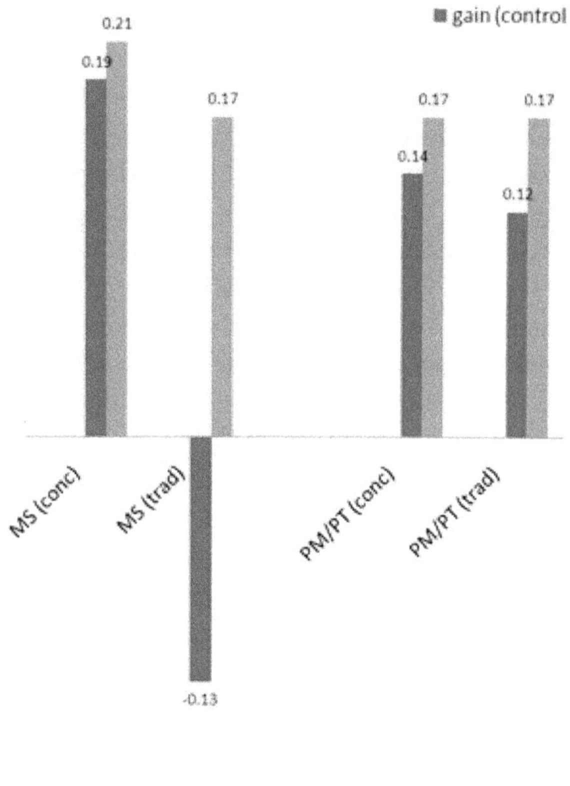

Figure 2. Comparison of net-learning gains by content area and item type II

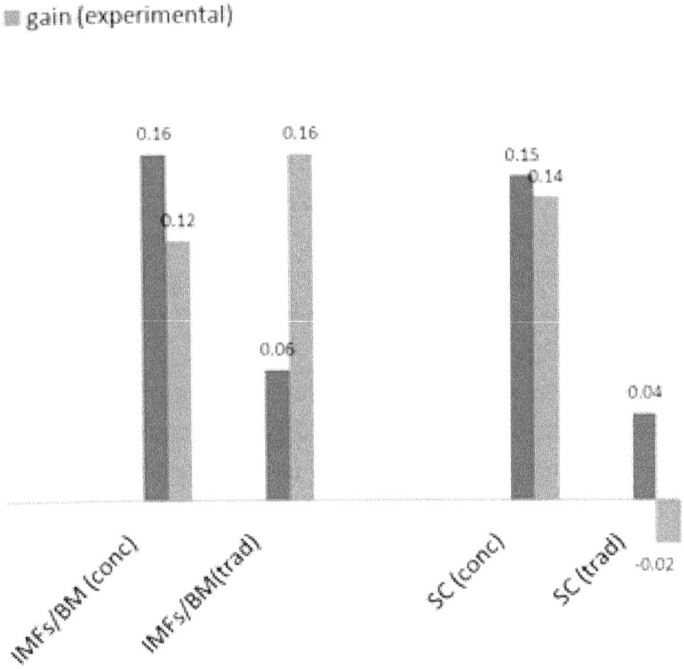

Comparison of Net-Learning Gains by Content Area and Content Type I

Figures 1 and 2 show results comparing net-learning gains by content area and item type in a two sections of a general chemistry course with the experimental class learning and using concept maps (Tang, 2013). Both figures show net-learning gains associated with non-use and use of concept mapping in two sections chemistry courses. The blue bars represent performance gains in the control classroom (no maps used) and the green bars represent performance gains in the experimental classroom (maps were used). A key to the chemistry content areas and types of items is included in between the two figures. The numerical values represent the average change in item difficulty for each content area. For example, a value of 0.19 means that an average of 19% more students answered those items correctly on the post test. Negative values indicate a decrease in average student performance from pre to post assessment.

Examination of the two charts show that the experimental class with students learning and using visual learning concept maps (vLms) outgained the control group in both concept and traditional quantitative areas with the exception of sections on the conceptual understanding of solution chemistry and intermolecular forces and the behavior of matter. The control group show significant gains in both the conceptual and quantitative aspects of molecular structure, properties of matter and the periodic table – all sections that involves spatial reasoning – and the quantitative section of solutions chemistry. When the experimental group applied visual learning maps to the textbook section of molecular structure, the mapping group performed a full 70% better on questions than the control group/

The results from the study also showed that more students in the lower ability level (lower 3rd of the class) made greater transitions into the higher abilities levels in the experimental class as compared to the higher level in achievement students. (Tang, 2013).

In a second experiment, with STS social science students UCD investigators used pre and post-tests, different versions of the ACCUPLACER (Mattern, 2009) to determine if the students increased their reading comprehension before and after being taught how to use the Visual Logic Maps. After a 1-tailed t-Test (paired two sample for means analysis,) the data revealed that the mean difference between pre and post reading efficiency scores was statistically significance at the 0.05 level ($0.048 < 0.05$ with a 95% confidence level). Because of these results, it is quite reasonable to hypothesize future student use of visual logic conceptual maps as an intervention tool on the undergraduate level would further improve STEM learning (Tang, 2013).

Visual Logic Maps

In addition, the foregoing experiments show strong evidence that reading effectiveness, spatial thinking and concept mapping are interrelated. It is therefore further hypothesized that the affinity of Visual Logic Maps (vLms) to spatial or structural learning may be due to the origins of verbal concept maps in the syntax of sentences which are parts of thoughts, which is heart of the cognitive processes. Figure 3 is a concept map conflated with a traditional grammatical example of diagramming of a sentence awhile the next figure is a visual logic map a kind of concept map that will be discussed in greater detail further on in this chapter. Both maps are presented as generic conceptual maps to visually illustrate the analogy: diagramming sentences is to grammar as concept mapping is to concepts. What will be argued for the rest of the chapter is that although Visual Logic Maps may find its origins in the verbal maps of Ramus and Novak, they have evolved into maps that have more in common with spatial maps found in specific domains such as logic and organic chemistry.

A. Sample Conceptual Map of a Diagrammed Sentence (Figure 3)
B. Sample Conceptual Maps of Two Visual Logic Maps (vLms) (Figure 4)

Figure 3. Conceptual map as diagrammed sentence from a text as to why one team won and another lost

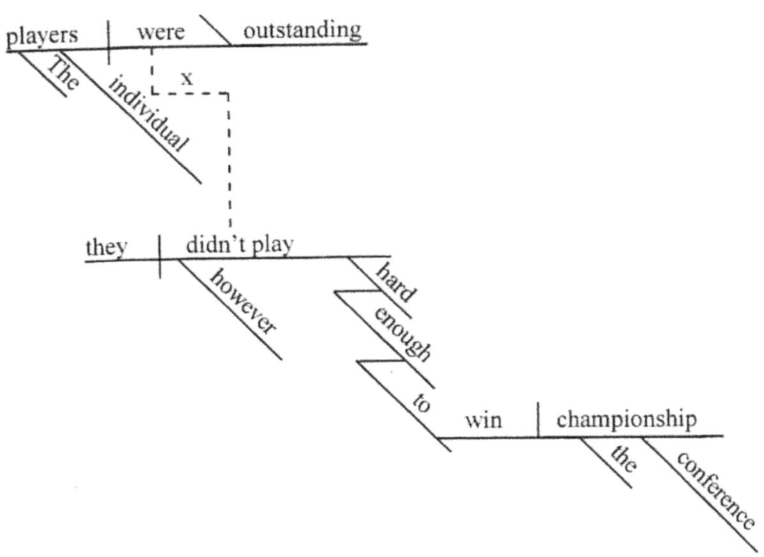

Figure 4. Visual Logic conceptual maps

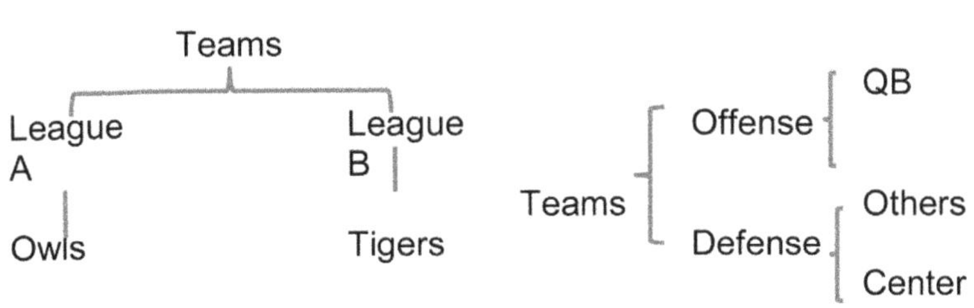

Combining Figure 3 with Figure 4 the following analogy can be made: "just as in English classes, diagrams are used to parse sentences to determine the meaning of sentences, vLms are used to analyze text or groups of sentences such as an article in a sports magazine to determine the articles meaning."

Now the reader is asked to look at both maps A and B and imagine unary symbols such as the letters of the alphabet and substitute words with letters. Once that is done, focus not on the letters or the nodes of the two maps but on their structure and a second analogy emerges: "grammar is to the structure of a sentence diagram as meaning is to the structure of a Visual Logic Map." Here we have a major distinction between visual spatial maps as exemplified by A as a diagrammed sentence and B as the concept mapping of a text on sports teams: In spatial maps, the glyphs and structure of the map is more important or at least equal in importance as words or the symbols that label those words. What follows is to demonstrate how different kinds of Visual Logic Maps can be distinguished from each other by explicit glyphs or symbols and the placement of the glyphs in a frame, that is by their structure rather than by verbal labels. In the vLms system, every position or shape, such as a line and its placement, (------) and arrow --→ or a text box [a] is meaningful, because they are designated symbols to indicate a particular operation. For instance, a line without a box indicates a categorization operation while one with a box indicates a properties map, while an arrow signifies an operation in time, or process. We know that Figure 3 is an analog map, to be discussed later in this chapter, because of the placement of terms which is a visual representation of the relation: "transformational grammar is to sentences as vLms is to text." We can now proceed to the other four glyphs and maps of the visual logic mapping system. It is important to note, however that both Thinking Maps and the Visual Logic Maps have evolved from their semantic origins and have developed into concept maps that are spatial and topological if not mathematical.

Consequently, the major difference between Visual Logic Maps and other concept maps, including Hyerle's maps, is in their construction and focus, where the new maps emphasize the symbols referring to connections and relations instead of the nodes, where Novak and other verbal maps place their emphasis. Again, another way of emphasizing the difference is in the Visual Logic Maps the linking symbols or fixed or constant while the nodes are variables. In Novak Maps and their cousins, both relational links and the nodes are treated as variables, while in the vLms, it is the logical glyphs that are treated as symbolic constants, while the nodes act as dependent variables which are usually labelled. The links or symbolic constants are referred to as glyphs that represent operations or functions that prompt the user to ask and answer seven logically questions to connect nodes as ideas in a meaningful and logical way mentally and visually.

The seven questions the glyphs prompt the user and constructor of the Visual Logic Maps are:

Quadrivium

- What is this a kind of and what are the kinds of this (*classification*)?
- What is this a part of and what are the parts of this (*structure*)?
- What is this a quality of and what are the qualities of this (*attribute*)?
- What is this a stage of and what are the stages of this (*operation*)?

Trivium

- What is this like and what are things or ideas are like this (*analogy*)?
- What does this connect to and what connects to this (*link*)? and
- What is this a context of and what are contexts of this (*frame*)?

Each of these logical prompts finds a corresponding statement in formal logic as shown in the figure and as such can be linked logically with other maps with a Boolean symbol such as =, ^ intersect. Once students have mastered

Figure 5. Table of Visual Logic Glyphs, the maps are designed to construct the meaning of the glyphs as logical statements in natural language

Glyph	Map	Logic Statement	Natural Language
	Classification	A= {a, b, c . . .}	a, b, c are kinds of A
	Attribute or properties of	∃x (Px): *x* is even.	P has the attribute of "even"
{	Structure	B = { . . .x, y. z}	x, y, z are parts of B
➔	Operation	A ➔ B	If A then B
+	Analog	A:B :: B: C	A is to B as B is to C
[ ]	Frame	A = Ø v { }	A is empty

the use of the glyphs as mathematical symbols they can begin constructing maps as an analytical tool to better understand STEM concepts, for which the Visual Logic Maps are especially suitable for reading because of heavy dependence on logic in scientific and technological texts. We now undertake to describe these maps in detail.

Classification Maps

When reading a biology text, the reading will frequently cover concepts such as structure as in anatomy or taxonomy as in a chapter on genus and species, where taxnomies can easily be organized by a classification map as can be seen in Figure 6.

Figure 7 is a more abstract version of Figure 6, where A = {B, C}, B = {F, G} etc.

Classification maps are hierachical trees and its nodes are separated according to genus-species or parent-child relations. Biology students will immediate see how classification analysis using the UCD concept maps can be helpful with learning concepts, which are aquired, not by rote memory from reading a text but by aquiring meaning, in this case the logical structure of taxonomies by using a classification map for analysis and reconstruction of material from a textbook.

Figure 6. Classification maps ask and answer the question: "What is this a kind of and what are the kinds of this?" i.e. Prokaryotic is a kind of living organism, etc

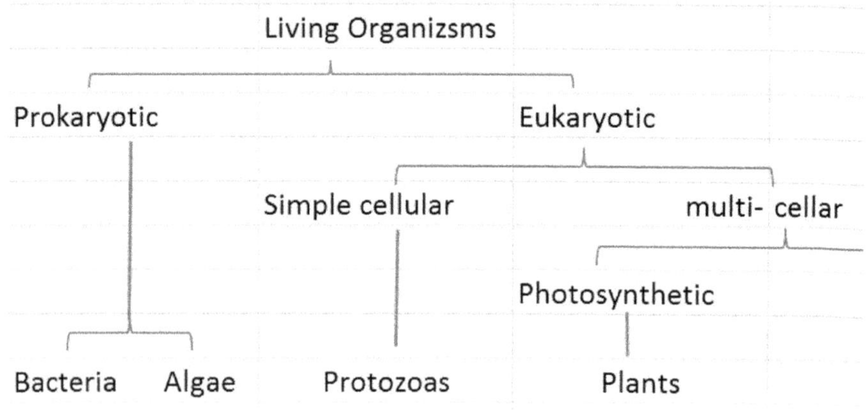

Figure 7. Non-verbal, more general version of Figure 6

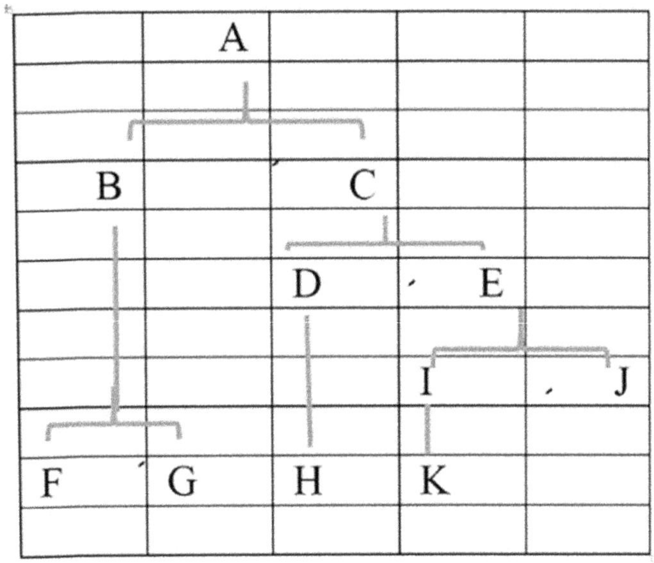

To take a further example, students encounter the following table that has been abstracted from a more specitic table where "Order" may be any biological order such as "Primates," which includes the family of "Hominoids" and so on down the line to the species, "Homo Sapiens," which is us. Abstracting these concepts the student may arrive at Figure 8.

Students adept at spatial reasoning, then might rotate Figure 8 mentally to arrive at Figure 9.

Figure 8. A generalized classification abstracted from of the order "Primates"

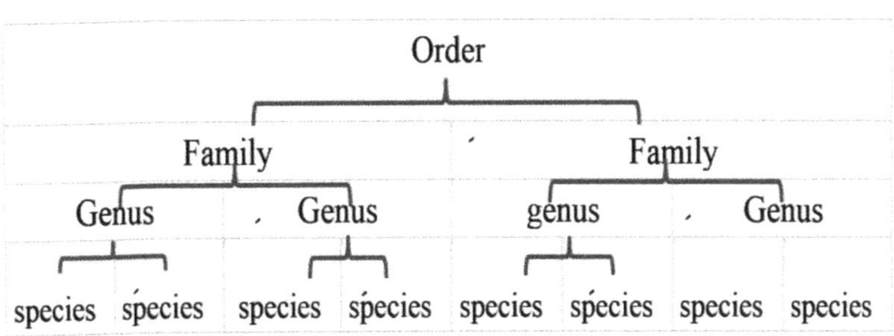

Figure 9. A spatial rotation of the concept map in Figure 8

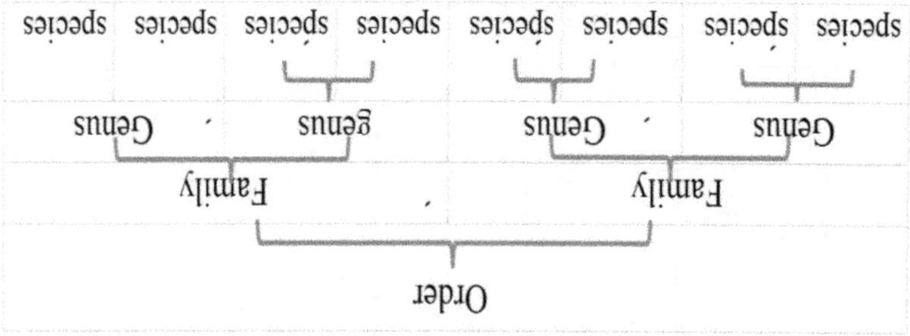

From this step, the spatially oriented students could reach the biological concept of adaptive radiation by changing the static classification mental schema into a developmental or evolutionary concept map.

Structural Analysis

The structural questions Visual Logic Maps ask are, "what is this a part of and what are the parts of this?" For example, if an instructor asked what is a key board a part of, the student would answer, "It is part of a computer." Then if the instructor asked what are the other parts of a computer, the student would answer, "The main parts of the computer, besides the keyboard, are the monitor and the computer itself, which has internal parts such as the motherboard which in turn consists of drivers, ports and the like."

The structural relation of parts, as in the case of the genus species or parent child relations of the classification maps is indicated by the same glyph, but instead of the glyph being horizontal it is vertical, as in Figure 10. Note again,

Figure 10. Parts map of Bohr Atom

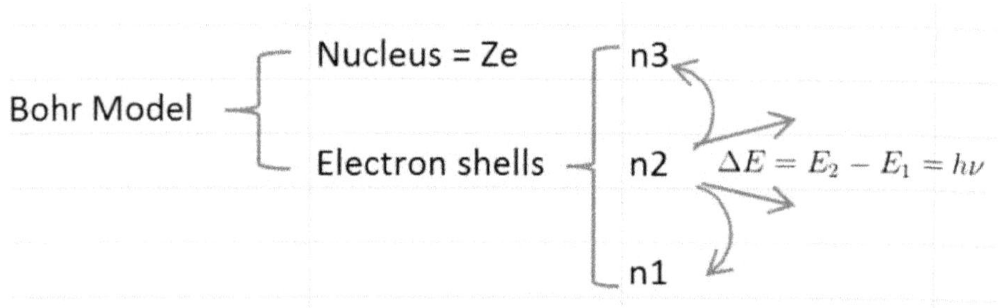

Figure 11. Qualities map of Bohr Atom with different glyphs and nodes

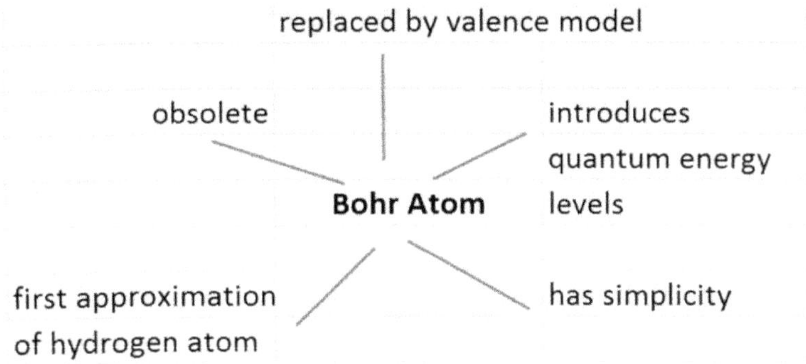

Figure 12. Two qualities maps on the "properties" of an apple. The Object Map is on the left and the Visual Logic Map, a generalization and abstraction of the Object Map is to the right

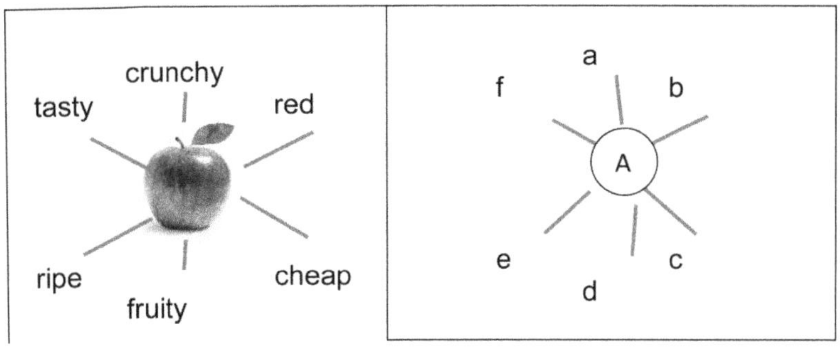

that a major difference between the object map and the vLm map, as in the case of the difference between Visual Logic Maps and Novak verbal maps, is the shift of emphasis from the nodes, which are the parts themselves, to the relational link which indicate parts, parts of parts and their relationship whole. This shift in gestalt is very clear when looks at the abstract vlm map to the right of the same figure. In essence what is being described here is very simple and intuitive, but the power of the glyphs is that they make explicit what was formerly implicit and as non-verbal abstract symbols they free the thinking process not only from the verbal to the non-verbal but from thinking patterns conditioned by the verbal. In this shift Visual Logic Maps are much more mathematical in nature than both Object and Verbal maps. All three maps, Object, Verbal and Spatial, are logical, but the logic among the three different types of maps are expressed differently. As in the case of classification maps, the nodes, which are variables, when labelled are represented as nouns.

Let us now say that now a chemistry reading assignment is on the Bohr atom. While reading the textbook introducing the concept students usually notice are the parts of the Rutherford-Bohr atom, which can be described by a structural concept map easily as in Figure 7. Note the parts maps, like the classification map, uses the bracket ({), but where in the first map the bracket is horizontal in the structural map the glyph is vertical. In Novak maps, which are highly verbal and both the nodes and connection symbols, usually arrows pointed down, the position of the arrow is relatively unimportant as it only indicates direction but other relation. In visual logic maps, the position of the connecting symbol is of utmost importance the position, whether it is vertical or horizontal indicates and a completely different kind of relationship.

Qualities Maps

Whereas structural maps are associated with parts and things, visual logic qualities maps are associated the properties, characteristics or qualities of things and concepts, and is indicated non-verbally by the simple glyph, the straight line. The qualities glyph indicates the different properties often as radiating lines from a central node such as in the in the properties of an apple.

From our experience, students often confuse structural maps with classification maps and confuse structural maps with qualities maps. A classification map might clarify this ambuiguity.

In Figure 13, note that the classification glyph, (horizontal bracket) produces a classification map (A, left), while a structural glyph (vertical bracket) is

Figure 13. A taxonomy of Visual Logic Maps and Glyphs

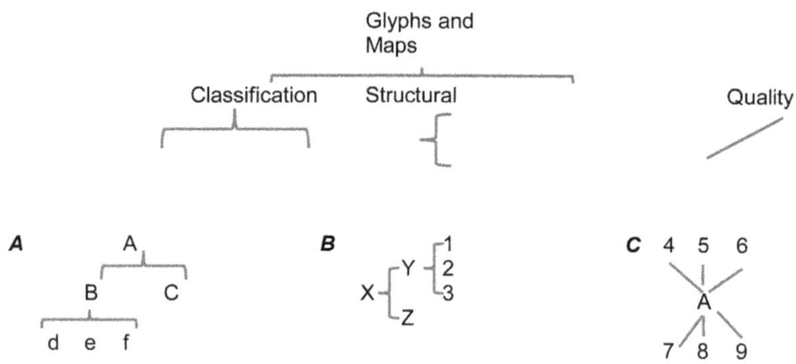

used to construct a structural map (center, B) and the line glyph is used to construct a qualities or properties map.

Operational Maps

Classification, structural and qualities maps are ontological maps, meaning that they are about "being," associated with the verb "to be. ontogeny is a fancy word for "what is." Therefore, the three maps previously discussed are categorized as static maps as they are unchanging pictures of things and concepts Ontogeny maps that change in the Visual Logic Mapping systems are called operation maps, which deals with stages, steps, processes, evolution, devolution, revolving, cycles and the like. Most simply. they take the form of lists, but to make things clearer, we use the arrow glyph. Generally, an operation map asks and answers the question, "What is this a stage of and what are stages of this?"

What is discussed here is a transformation rule wherein a classification chart is changed into an operational chart as previously a classification chart had been changed into a qualities chart.

Figure 16 in its most sophisticated and powerful form, an operation formula, as opposed to a map, takes the form A -→ B, which in formal logic states, "if A then B" ---- if A then B ^ C. If B then (D v D v F).

In many ways Operation Maps are the most familiar, because it is a linear way of organization conditioned by the text media. A list is an Operation Map if it organizes information according to stages or phases. It asks the question, "What is this a stage of and what are the stages of this? ---a question, as in the case of all the root questions, can be reduced to the statement: x(Ax ^ Bx) and the "if – then" statement. If A then (1,2,3,4) but if B then (5,6,7,8)

Figure 14. A static classification chart

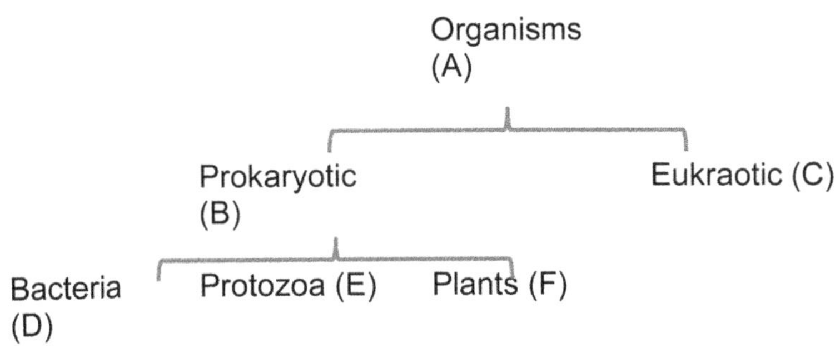

Figure 15. A dynamic operational map

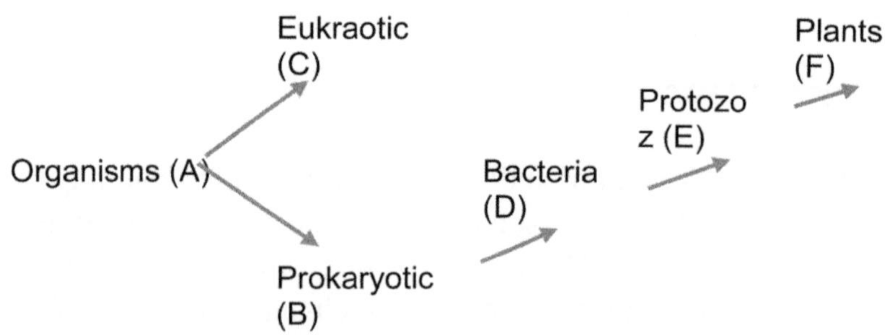

Figure 16. A complex dynamic operation map

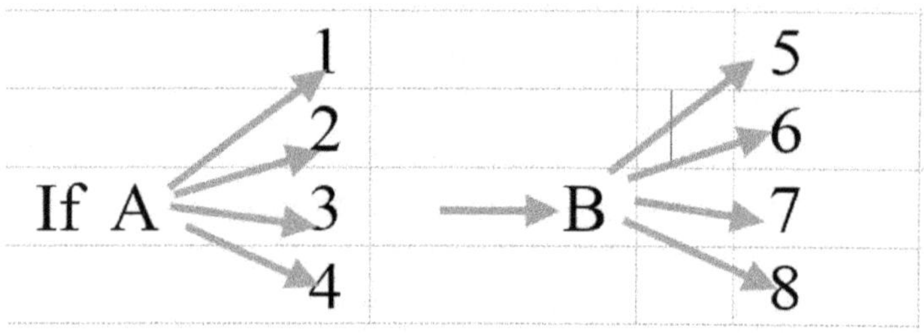

The "If – Then" statement as expressed in Figure 16 uses facts, rules and definitions to deductively reach a conclusion. For example, take the statement, "If you get good grades then you will get into a good college": the part after the "if": you get good grades - is called a hypotheses and the part after the "then" - you will get into a good college - is called a conclusion. In other words an if-then or conditional statement is simply a hypotheses followed by a conclusion, and is at the heart of all scientific and engineering practice.

Operation maps may appear rather simple and elementary but logical complexity is built on the simple and elementary as is shown by the recursion diagram in Figure 17. Recursion can be defined as a process of defining and redefining a function or calculating a number by the repeated application of an algorithm in a cycle or spiral in which each cycle is repeated again and again, with each repetition feeding on and modifying itself with continuously new input (Dijkstra, 1960). One of the most powerful aspects of language is its recursive nature as shown in Figure 17, from which a person can say, "Alice eats and runs quickly [while] Bob eats and runs slowly [although when] Alice eats slowly [she] runs quickly" and on and on through several iterations (OText, 2016).

Figure 17. A recursive operation map showing recursion in speech. Image source: https://www.otexts.org/node/834

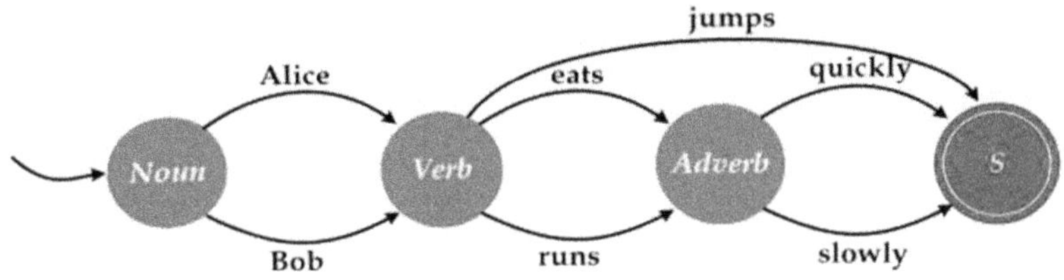

Analogical Maps

An analogical map has already been encounter in the previous chapter where an analogy was made between the Genesis version of creation and the Great Bang theory in the context of illustrating the scaffolding and bridging aspects of advance schemas to help student learn by constructing their own knowledge from previous knowledge. The analogical glyph used to construct analogical maps is represented by the simple cross. Although it is usually much larger than the other glyphs and operations in the background. Analogies are based on the logical relation called similitude, where, a:b:: c:d or a:b ≈ cd. The root glyph, **≈**, means, "is like something else."

Analogies and metaphors are more complex forms of a similitude. Rigorous examples of analogies are those found in the Miller Analogies which often appear in standardized tests, especially for graduate school admissions as analogies are purported to measure higher level thinking. Miller Analogies have become a favorite testing item to measure aptitude in computer science and analogies are basically a more complex form of similitudes, or correspondences as in A ≈ B, A corresponds to B or A is similar to B. Analogy Maps are two dimensional presentations of the relationship: A is to B as C is to D (Sternberg, 1983). For example, in Figure 18, painting is similar to music in that both are art forms and in the arts, Mozart is to music as Monet is to painting.

As in the case of operation maps becoming recursive maps, the simplicity of analogical maps belies its sophistication for higher order thinking. For example, a biology student might come across the following text.

The theory of recapitulation, also called the biogenetic law or embryological parallelism—and often expressed as "ontogeny recapitulates phylogeny"—is a disproven biological hypothesis that in developing from embryo to adult,

Figure 18. Two analogy maps, one in the arts (left) and the other on evolution (right)

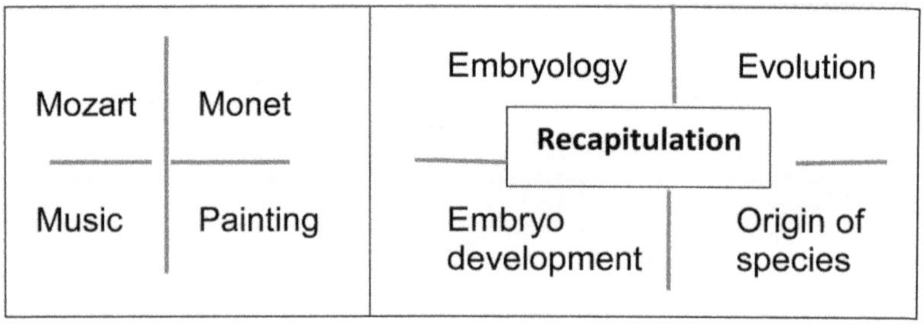

animals go through stages resembling or representing successive stages in the evolution of their remote ancestors. With different formulations, such ideas have been applied and extended to several fields and areas, including the origin of language, biology, cognition and mental activities, anthropology, education theory and developmental psychology. While some examples of embryonic stages show that superficial features of ancestral organisms exist, the theory of recapitulation itself has been completely disproven within the field of biology.] By contrast, there is no consensus against its validity outside of biology; recapitulation theory is still considered plausible and applied by some researchers in fields like Behavioral Development, the study of the origin of language, and others. Source: https://en.wikipedia.org/wiki/Recapitulation_theory.

Verbal Analogies can come in many types as illustrated by the concept map in Figure 19.

Moreover, they need not be verbal as exemplified in Figure 20.

To summarize: An analogy is a noun which signifies a comparison between two things, typically for the purpose of explanation, clarification, or style enhancement. For example, "an analogy exists between the workings of nature and those of human societies." Another example would be, based on .a correspondence or partial similarity as in "the syndrome is called deep dysgraphia because of its analogy to deep dyslexia," or to compare a thing to something else in significant respects such as, "works of art were seen as an analogy for works of nature." Analogies which are restricted to the verbal form are called "similes" which can be defined as figures of speech involving the comparison of one thing with another thing of a different kind, used to make a description more emphatic or vivid such as "he was brave as a lion and crazy like a fox". Analogies, however, can also be non-verbal as exemplified in Figure 20.

Figure 19. Common Types of analogies. http://cursa.ihmc.us/rid=1180035154109_133022099_36801/Common%20Types%20of%20Analogies.cmap

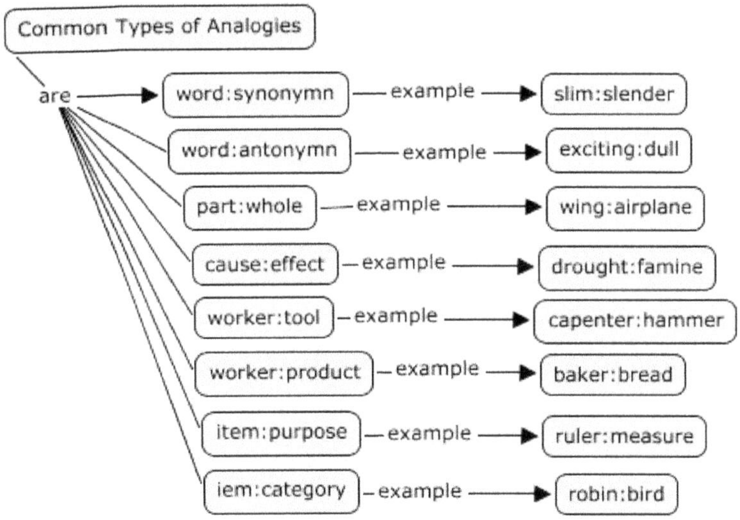

Figure 20. Non-verbal spatial analogies

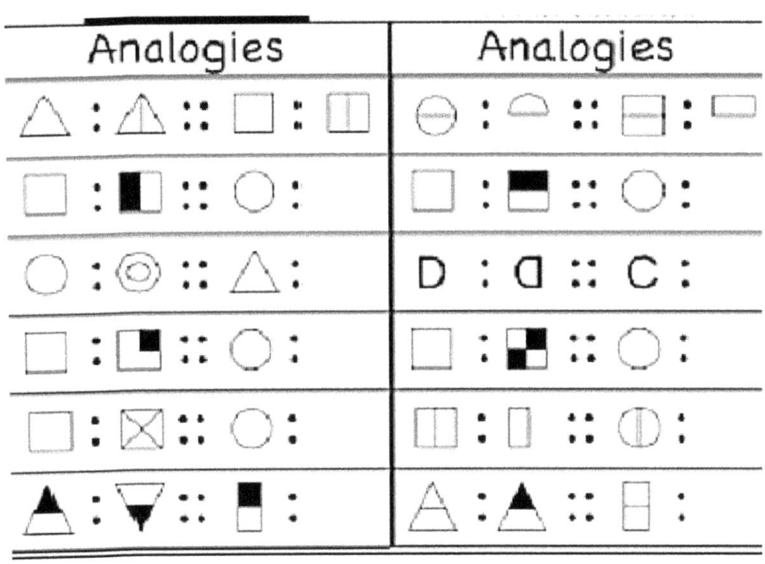

Metaphors, similar to analogies and similes are figures of speech in which a word or phrase is applied to an object or action to which it is not literally applicable, such as "“I had fallen through a trapdoor of depression,” said Mark, who was fond of theatrical metaphors." For example, Mark could literally be “crazy like a fox,” could not really fall through a “trapdoor of depression.”

Frames

Frames are used to give the maps a context, serve as a space for glosses and to clarify ambiguities. It functions as a structure that provides the background or other information to help users understand or interpret the elements contained in root maps. Frames within frames can also be used for brainstorming or as a preliminary and summative way to organize information. In using a similar map in this way we found that middle school children actually discovered the excitement of meta-cognition when working on a STEM project classifying and analyzing rocks for a geology assignment (Tang, 2012). Link Maps connect words or images to another as hyperlink connects words and images on the Web. After reducing complex text to the simple maps, we can connect these simpler units into the complex via the link. This is where the creativity and magic happens.

A frame is a structural system that supports other components of a physical or abstract construction and its glyph in Visual Logic Maps is represented by an empty box, and like the analogy cross glyph can take up most of the page, and therefore is usually a much larger glyph the others in the system. Conceptually it is a container that holds concepts and a container can be contained in a larger container or contain smaller containers in a nested arrangement. Frames can represent frames in the real world. For example, in engineering, a frame, a basic structure is designed to bear a load in a lightweight economical manner. In the social sciences, framing is a terminology used in communication theory and sociology, where it relates to the contextual presentation of media content. In literature, a frame story (also frame tale, frame narrative, etc.) is a literary technique that sometimes serves as a companion piece to a story within a story (Richardson, 2002). The story frame leads readers from a first larger story or narrative into another, smaller one (or several ones) within it (Hyerle, 2011. In mathematics, an ordered basis is called a “frame” and a frame bundle is a principal fiber bundle associated with any vector bundle (Ferreira, 199). A Frame of reference is the set of coordinates in which a system is observed and a frame of a vector space, is a generalization of a basis to sets of possibly linearly dependent vectors which also satisfy the frame condition. Staying within the domain of computer science, a GUI is a frame used to hold other widgets in a graphical user interface and in networking a meta-frame is a data transmission unit or network packet that includes frame synchronization information. Framing on the World Wide Web is a method of displaying multiple HTML documents on one page of a web browser (Embley, 1980; Connallen, 1999).

The frame could be considered as a device to help the user organize information easier. For example, Figure 21 is a frame map anchored the number 10 and surrounded by different ways the number can manifest itself. More importantly, however, the frame functions as a map or structure that provides the background or other information to help users understand or interpret the elements contained in root maps. The frame provides the context or matrix to give clearer meaning to any of the six maps that may be ambiguous. For example, in Figure 21 "card games" is the frame for drawing the "bridge" map, where the term "bridge," may be confused with an engineering structure where the main purposes would be completely different than those of the card game. An analysis of an engineering bridge would be better served initially with a Parts Maps consisting of beams, trusses and arches.

In Figure 21, a simple frame, labeled "ways to make 10" was drawn by an eighth grader (right) and the characteristics of bridge is contained in a frame labelled "Card Games" to clarity that trick taking, bidding, scoring and

Figure 21. Two frames, one for "ways to make 10" (right: https://www.pinterest.com/edtechnerdo/thinking-maps/) and one to clarify that bridge as a kind of card game.(left) and not a structure to span a river or another gaps in the terrain

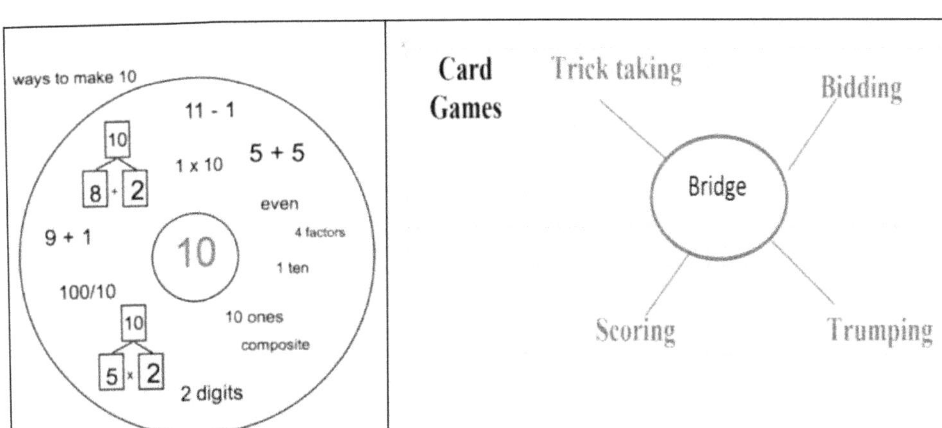

trumping are characteristics of bridge that has nothing to do with a bridge in engineering.

The nested container function of frames, however, is where the frame maps may have their greatest potential as exemplified by container or boundary mathematics, which take un-labelled glyphs and their subsequent maps or spatial equations to unprecedented levels of non-verbal and non-linear abstraction. For example, the mathematician, William Bricken (2005), teaches his students a form of spatial logic to challenge the binary basis of logic represented by the symbols, 0 and 1, "True" and "False" and "Yes" and "No" by asking the question: "Is there a simpler approach? Can logic be expressed in a unary notation, with only one value?" According to Bricken the answer is "yes" to both questions and demonstrates how through a system of math that uses delimiting tokens, or containers, as both constants and functions based on Spencer-Brown's logic as found in his *Laws of Form* (1972).

Spencer-Brown's spatial logic is based on two very fundamental statements that are expressed as follows: The Law of Crossing: (()) = and the Law of Calling: () () = (), which can be translated as follows if we call the symbol () a frame: 1) A frame contained within another frame crosses into absolute void (i.e. null void) and 2) a frame () called again as an empty frame remains the same, an empty frame, or a relative void. Using the empty box: as the glyph for a frame, we get frame maps of the two expressions found in Figure 22.

With Bricken's boundary math, frames become the glyphs to express formal logical laws in the highest abstract form possible, with not only no words, but not even labels such as A, B C, and X and Y to orient the reader or user. The point here, as in the case of the discussion of the other glyphs from classification to the if-then statement, extremely simple visual logic glyphs to construct extremely simple maps can suddenly be transformed

Figure 22. Two primitive laws, the Law of Crossing (left) and the Law of Calling (right) from Spencer-Brown's, Laws of Form

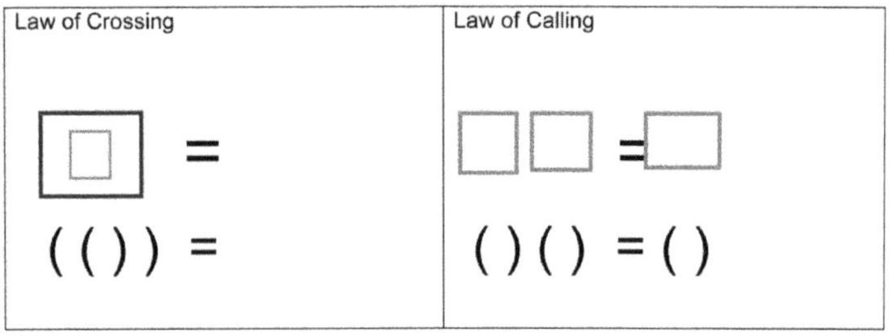

into the highest and extremely complex non-verbal abstractions possible (Bricken, 2005).

The Link

What connects one word or image to another on the Web is the hyperlink, and this same link when constructing more complex vLms connects one map to another by similitude or association. After reducing complex text to the simple maps, we can connect these simpler units into the complex via the link, where the creativity and magic happens. In Figure 23 note that our complex map consists of different logical maps mixed together as well as linked together.

Ideally mappers should be able to build a complex system of maps using the link. One possibility would be link the different two dimensional maps into al structure resembling a Geographic Information System (GIS) relational database. Ideally a student could build a complex system of maps using the link. One possibility would be link the different two dimensional maps into al structure resembling a Geographic Information System (GIS) relational database.

Figure 23 consists of three maps, a parts map of the nucleus, consisting of inner and outer orbits; an analogy map, that states, atom: solar system:: nucleus: sun; and a second parts map stating that the solar system consists of the sun and planets, which consists of the inner and out planets.

Complex Maps

Classification, structural and qualities maps can be combined to make more complex maps. For example, an instructor might assign students to analyze the following passage using two kinds of maps, Qualities Maps and Structural (Parts) Maps.

Waste (excretion): this term refers to removal from the body of waste products that may have been formed by metabolic processes. Examples include: 1) Breathing out (of the body) carbon dioxide (C2O) and water (H2O), and the waste products from aerobic respiration; 2 sweating, which is the removal of waste products from the body via the skin; 3 urinating. (Urea is nitrogenous waste, i.e. it is from proteins; and defecating, which is defines strictly as

Figure 23. Two parts maps (nucleus and sun) linked to an analogy map (atom: solar system:: nucleus: sun)

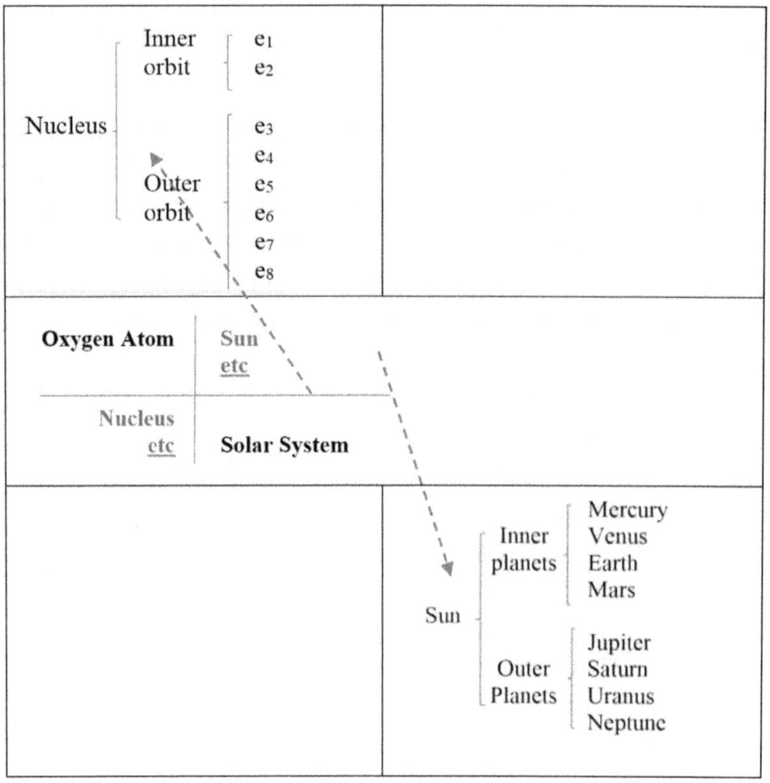

materials that leave the body by the process of defecation are indigestible food materials - rather than waste products formed by the metabolic processes that occur in the body. Their color is due to the indigestible food itself, and to the effects of any excess bile.

The structure of a student complex map may look as follows:

Figure 24 shows three maps: map M, which has the properties of A, B, C, D, with a parts map consisting of parts, 1, 2 and 3, which is linked to an O map, which has the properties, D, E, and F.

To summarize classification glyphs and maps ask and answer the question, "what is this a kind of and what are the kinds of this? Classificatory relations are best expressed as genus species or parent tree relations and generally hierarchical in structurer. Structural maps and glyphs deal with material parts

Figure 24. A complex Visual Logic Map consisting of two Qualities Maps (M ^ O) and a structural map (B)

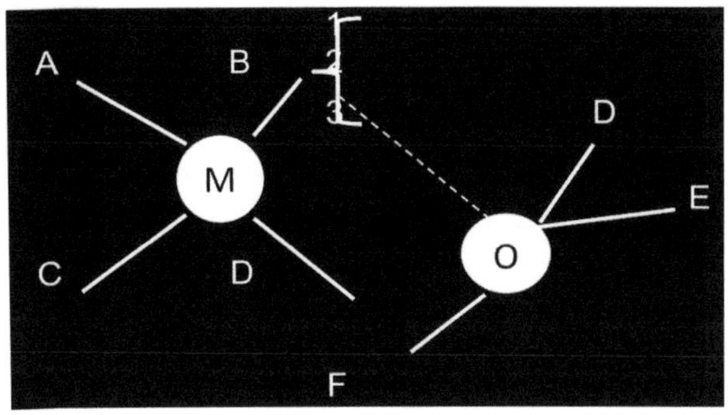

and wholes and asks and answers the question, "what is this a part of and what are the parts of this (structure/whole)"? Qualities maps indicate properties and characteristics that belong to a thing or concept and are represented as lines radiating from a central node attached to nodes indicating specific characteristics and the like, asking and answering the question, "what are the qualities, properties, etc. of this and what thing, event, etc. has this quality?" Structural maps and qualities maps have similar structures and can easily be derived from classification maps.

Operational maps can easily be transformed from classification maps as well and poses the question, "what is this a stage of and what are stages of this?", while analogy maps ask, "what is this like and what are the likes of this?" The links in a concept map and concept maps of course link concepts to as well as maps to each other. Frames, give concept maps context. All of the maps can be reduced to the statement A = {a, b, c, d . . .} in set theory or $\forall x((x \in S \wedge x \neq a) \rightarrow \cdots)\forall x((x \in S \wedge x \neq a) \rightarrow \cdots)$ in first order logic or the predicate calculus.

CONCLUSION

The following is a table that compares the three concept maps analyzed as a taxonomy of Object Maps, Verbal Maps and Spatial Maps, after Blazhenkova and Kozhevnikov's three cognitive styles (2009) and M. J. Eppler's classification of concept maps (2006).

Table 1.

Object Maps	Verbal Maps	Spatial Maps
An object map is a multi-colored and image centered radial diagram that represents semantic or other connections between portions of learned and other material hierarchically	A verbal (Novakian) concept map is a top-down diagram showing the relationships between concepts, including cross connections among concepts, and their manifestations	A systematic depiction of abstract concepts as a function of the entire map with an emphasis on links or connections over content nodes. Ideal for systems analysis.

Table 2. Function

Show sub-topics of a domain in a creative and seamless manner. Appeals to object learners, who tend to be non-verbal and non-spatial	Shows systematic relationships among sub-concepts relating to one main concept. Verbal types prefer this kind of mapping	Analyze a topic or situation through a systems framework as a function of entire map. Tends toward highly abstract, non-linear logical patterns

Table 3. Application

Object	Verbal	Spatial
Great for brainstorming, Mnemonics based on emotional part of the brain. Use as a pre-analytic intuitive device to stimulate new ideas. Right brain mapping.	Classroom teaching, self- study and revision text illustration, text analysis management discussions, personal note taking and reviewing; Left brain mapping	Non-verbal, symbolic representation of forms; tends towards specialization and esotericism; could serve as bridge between right and left brain functions

Martin J. Eppler is a Swiss organizational theorist, Professor of Media and Communication management at the University of St. Gallen, and director of its Institute for Media and Communication Management. Professor Eppler is internationally well known for his contributions in the field of knowledge management and concept mapping as an alternate visual means of cognitive processing and organization.

Eppler received his PhD in Management from the University of Geneva in 1988 and been a visiting professor at Georgia Tech in Atlanta, USA), Simon Fraser University in Vancouver, Canada, the Central University of Finance and Economics (CUFE) in Beijing, at the Aalto University in Finland and at the Universidad del Pacifico Graduate School in Peru. Professor Eppler has published numerous articles and books in the fields of organization studies, knowledge communication, and visualization.

THE FUTURE OF CONCEPT MAPPING

Along with the expected correspondence between cognitive styles and cognitive mapping this chapter's analysis also found --- by reading between the lines --- an apparent correspondence between the retrieval of concept mapping with the rise of the Internet and the World Wide Web. According to Okada (2008)...

We envisage that human and machine knowledge mapping will eventually converge. Software agents will work continuously in the background and on demand, generating maps and alerts that expose potentially significant patterns in discussions and publications (e.g. term clusters; hub nodes; pivotal papers; emerging research fronts; supporting/challenging evidence; candidate solutions). Analysts will assess, further annotate, and add new interpretive layers. . . The layers that analysts will add to machine generated maps will, therefore, also reflect the community's deliberations—whether in meetings or the literature— adding important connections and summaries that are not in the source documents/datasets. (p 10)

This finding of an apparent correspondence between concept mapping and the World Wide Web is a hypothesis worth investigating further especially in the context of the impact of communication media as technologies on our epistemologies (Tang, 2017).

In Part Two of this Insight Book, we developed a taxonomy of Visual Logic Maps and glyphs guided by and framed by O. Blazhenkova and M. Kozhevnikov's cognitive style theory which states that different thinking styles can be categorized in to object, verbal and spatial, with object thinkers having a preference for processing information through colorful, concrete, if not multi-sensoria schemas. Verbal thinkers appear to prefer processing information with schemas that are primary verbal, with the characteristics that go with that medium: i.e. abstract, linear, sequential and in general reductionist with words to construct sentences, with are considered to be the main parts of concepts. Spatial thinkers are like object thinkers in that both prefer to think in terms of non-verbal structures and patterns spatial cognition being more abstract and logical in the formal sense of the word than object thinkers. In the taxonomy delineated in Part II it can be inferred that object thinkers would prefer to use Buzan's colorful, freeflowing, concrete Mind Maps over the other two; that verbal thinkers would prefer to use Novak's concept maps; and spatial thinkers, Hyerle's Thinking Maps.

Parts I and II also attempt to make a clear distinction concept maps that are meta-maps in that they are interdisciplinary if not transdisciplinary and specific domain maps such as those used in organic chemistry or formal logical analysis. Another area that needs investigation is a thorough comparison between concept mapping of text material versus mapping the so called real world as best exemplified by G.I.S. maps, which are used to analyze the real world.

For example, as there are different kinds of G.I.S.maps such as the vector and the raster, there are different kinds of spatial maps, such as the Visual Logic Maps, discussed in great detail in this book. In this new research, it should be emphasized that where we use G.I.S. maps to analyze the real world, we use spatial maps according to our typology, mainly to analyze textual material, whether in hard text or in digital formal.

Analog to G.I.S. Mapping

As in the case of G.I.S. the spatial analysis of text, as exemplified by the vLm maps, follows strict logical rules that are the root are mathematical. It is this mathematical basis that make these maps different from other maps such as mind maps, verbal concept maps and other organizational graphs. Figure 25 shows how Visual Logic Maps, as a layered spatial maps, can be used to analyze textual material much as a geodetic scientist uses GIS maps to analyze terrains in the real world. A student, for example, would take a page of text, read that text and then draw different vLm maps from the root glyphs to construct and better understand the relations and connections among the different concepts found in the text.

Figure 26 shows how the Visual Logic Maps can be used to analyze text material as individual maps that extract meaning from textual material by abstracting from the text and laying out important concepts as spatial maps to help in processing textual information in non-linear and highly structured visual patterns.

In this concluding chapter we also detailed the correspondence between vLms glyphs and vLms maps in that the seven glyphs constitute seven different kinds of maps, including 1) classfication, 2) structural, 3) qualities, 4) operation, 5) analogy, 6) frame and 7) link maps.

Figure 27 shows four of the seven glyphs that are used to construct corresponding maps in that classification glyphs are used to construct classification maps, part glyphs, structural maps, qualities glyphs, quality

Figure 25. A G.I.S. map with layers that find their analog with Visual Logic spatial Maps

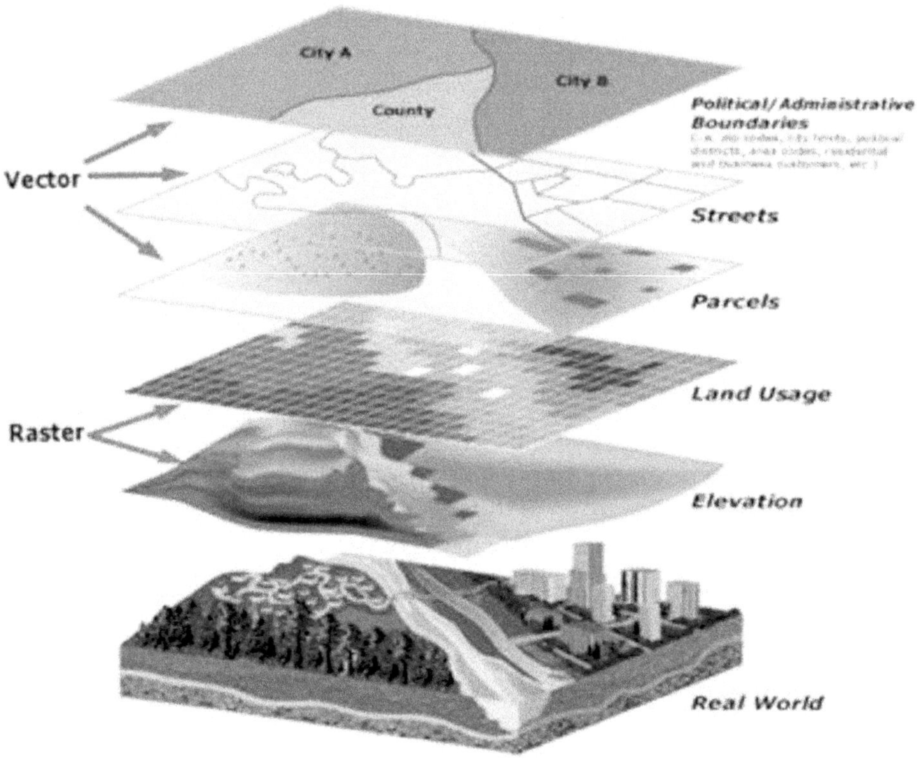

Figure 26. Graphic showing how Visual Logic Maps can be used to different lens to analyze text as GIS maps are used to analyze different layers of terrain in the real world

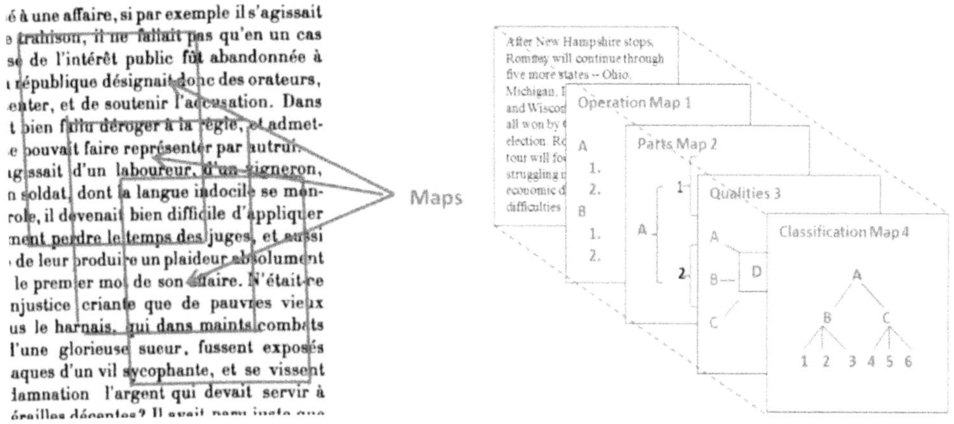

Figure 27. Four of the seven glyphs that are used to construct corresponding maps by the same name

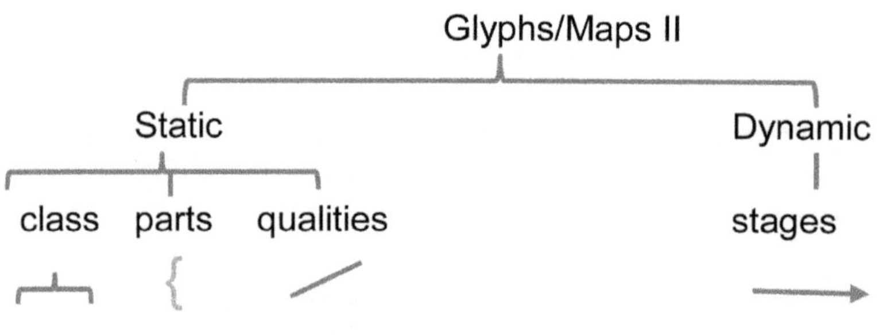

maps and stage glyphs, operational maps. Figure 28 show the remaining three maps, where the correspondence between glyph and map is not as definite as the first four glyphs/maps. Note also that of the four maps, only operational maps are dynamic.

Figure 28 gives the last three of the seven glyphs/maps, the link, frame and analogy, with the frame and analogy maps being static and the link dynamic, if it acts as an actual hyperlink connecting different maps in a Web or other computer environment.

Recapitulation

One of the analogies used in the book's on recapitulation from the "law" that ontogeny recapitulates phylogeny. Although this statement is not really a law, it summarizes neatly the historical trend of concept maps discussed in Part I with a general trend that is discernable from the current state of affairs with

Figure 28. The link, frame and analogy glyphs with semi-corresponding maps make up the rest of the seven glyphs/maps. Again, of the seven glyph/maps only operations are dynamic

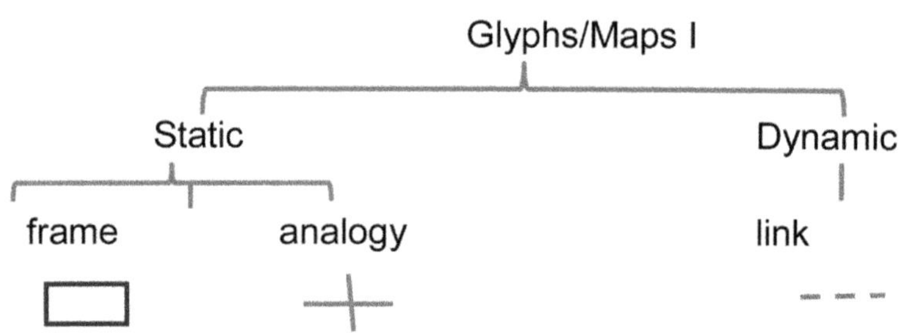

concept maps as discussed and analyzed in Parts II and III. Essentially that trend can be that concept maps appear to be developing from the object maps to verbal maps to spatial maps. In short the ontogeny of today's retrieved maps appear to recapitulate the historical maps, Ramon Lull's (object) to Peter Ramus' (verbal) to highly abstract, specialized logic maps of Leohard Euler and John Venn (spatial).

Figure 29 is an example of a highly abstract spatial/verbal map used in Web programming and is exemplified here to illustrate another aspect of concept maps that has only been implied: concept maps are only models of objects, whether the objects are textual or actual, and therefore only approximate what is being mapped. In fact as the general semanticist, Alfred Korzybski (1958) warned, "the map is not the terrain" and "the word is not the thing" for if they were, as analytical tools, they would be useless. With this caveat

Figure 29. An example of a highly abstract spatial/verbal map used in Web programming. Source: http://www.cs.swan.ac.uk/~csowen/InformationRealisationAppendix/ontologies/index.html

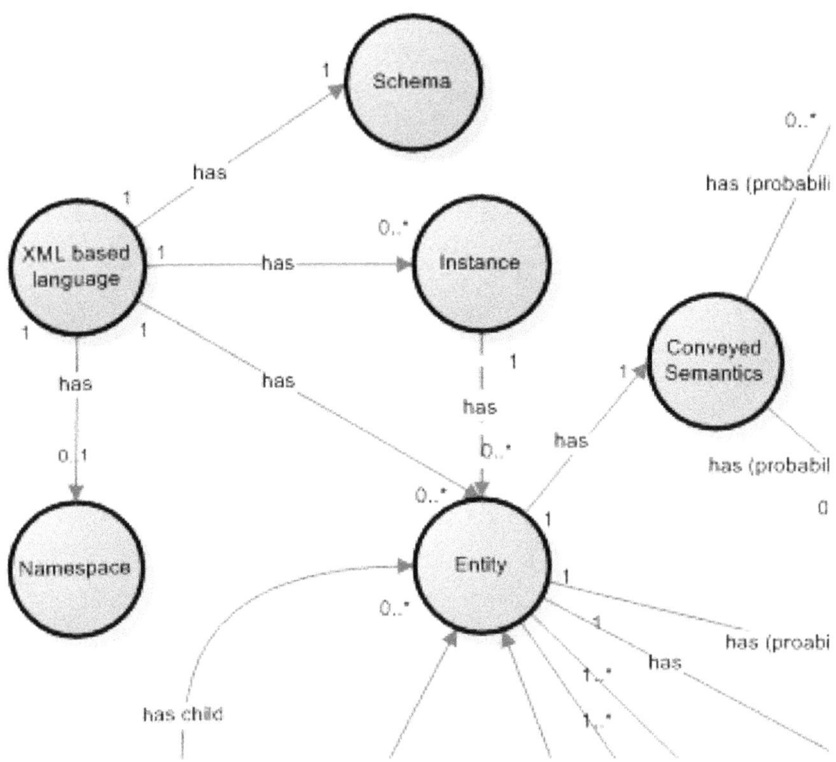

is argued, as for which concept map is the best, the answer is "it depends." The best map is the map that works best for the individual and the nature of the problem being solved through conceptual mapping. For example, Figure 30, looks at first if it were a Mind Map, but on closer examination, it probably is best classified as a Novakian concept map in spite of all the colorful parts, which appears to be quite distracting as presented this page. However, keeping in mind McLuhan's "law" in this context, "the medium is the message." That is Figure 30 is taken from a Web page concept charting software package that enables users to draw three dimensional maps with hyperlinks to create what we would call an object map.

Figure 30. A highly colorful Novakian Map copied from a Web page, explaining why on a static, abstract book page, it looks out of place. Image source: http://www.freewaregenius.com/map-your-mind-in-vibrant-3d-graphics-with-topicscape-student-edition/

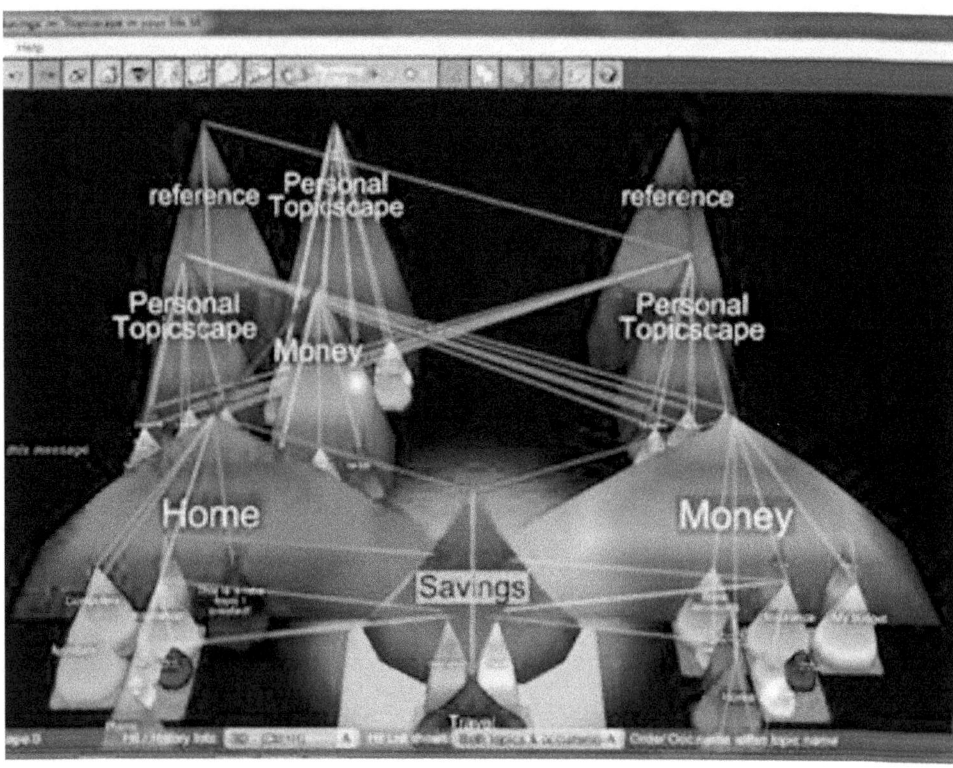

**For a more accurate representation of this figure, please see the electronic version.*

The Medium is the Message

Marshall McLuhan's aphorism "the medium is the message," also can be used to historically to explain that radical media changes will result in radical changes in all areas, including science and technology, not to mention science and engineering (STEM) education, questioning the very idea of linear change and trends as outline above in regard to object, verbal and spatial learning with object, verbal and spatial maps, albeit that apparent congruencies. In any event whether the trend that we are witnessing in the development of conceptual maps is linear, cyclical or recursive, one thing that appears to be occurring is the trend towards specialization is paralleled by a trend towards general or meta-maps at the same time as exemplified by the main concept maps discussed in this book whether Buzan, Novak, Hyerle or Vistual Logic Maps. These key maps, it has been argued are general tools for cognitive processing and therefore are meta-cognitive tools that will continue to have a profound effect on education, especially STEM education, well into the future. As to the question, which of these concept maps discussed in detail, is the preferred map, the answer, is "all of the above," meaning different maps should be used for different ends. For example, if a person wants an emotional response to a problem or issue, using a Mind Map may be the best choice. However, if the desire is for a highly abstract more intellectual answer to a problem, a spatial map would probably be preferred.

As indicated at the beginning of this book, Marshall McLuhan (1992) and this theory of "the medium is the message" has been part of what he called the "hidden ground," which this conclusion has briefly raised to the status of "figure" We now conclude by retrieving another aspect of McLuhan him from the ground to discuss the media guru's own conceptual map, the Tetrad, as figure, to interpret what has happened on this topic and what we think will occur in the near future.

The Tetrad

In *Laws of Media* published posthumously, Marshall McLuhan (1988) summarized on, with his son, Eric, their concepts about media in a concise tetrad of media effects as means of examining the effects on society of any technology or medium. They defined the tetrad as a meta-concept map designed to measure the impact of technologies on the society and at the same time on our cognition, an impact neglected by most McLuhan commentators and scholars. The tetrad in this regard is a means of explaining mental processes underlying the adoption of a technology/medium by dividing its effects into four categories and displaying them simultaneously. Moreover, McLuhan designed the tetrad as a pedagogical tool, phrasing his laws as questions with which to consider any medium, including the medium of the mind, which in this case, will be applied to concept mapping as to how this technology impacts how we think or process information.

In its most general form, the tetrad questions are:

- What does the medium enhance?
- What does the medium make obsolete?
- What does the medium retrieve that had been obsolesced earlier?
- What does the medium flip into when pushed to extremes?

Figure 31 is an example as to how the tetrad as a complex analogical concept map can be applied to a technology, movies that can be downloaded from the cloud.

Figure 31. Impact of movies online.
Source: http://3.bp.blogspot.com/_3Z9X0KT5wOM/TGHHVUX-I5I/AAAAAAAAAB4/KdRYMiWF7iY/s1600/Picture+13.png

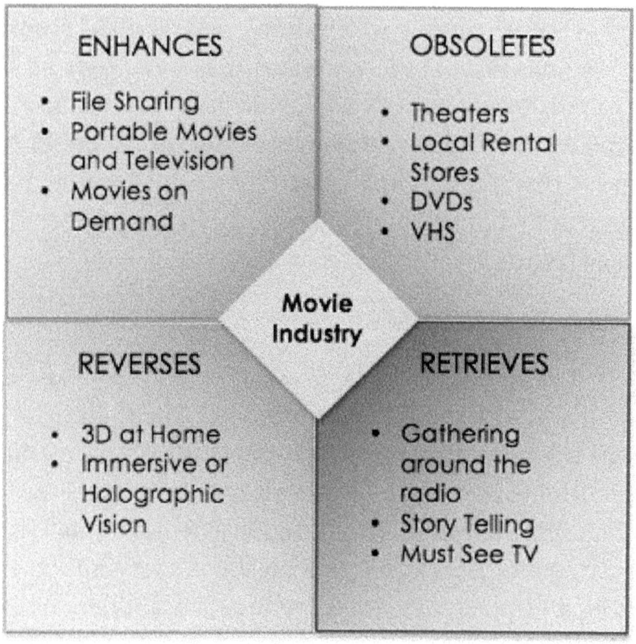

Accordingly, the ability to watch and download movies online will obsolesce theaters, local rental stores, DVDs and VHS, while the technology enhances file sharing, portable movies and television and movies on demand. At the same time the technology will retrieve gathering around the radio, story-telling and the multi-sensory experience. The same technology will flip into 3D at home, immersive holographic vision and virtual reality.

Applying the tetrad to a concept or idea as technology:

- Retrieves a concept from the ground
- Obsolesces other concepts and sends them back to the ground
- Enhances aspects of retrieved concepts to the foreground and
- Flips to (or reverses) the qualities of the concept into something else completely different and possibly even its complete opposite.

Which could result in the consequences mapped out in results as follows.

Figure 32. A tetrad analysis of concept maps

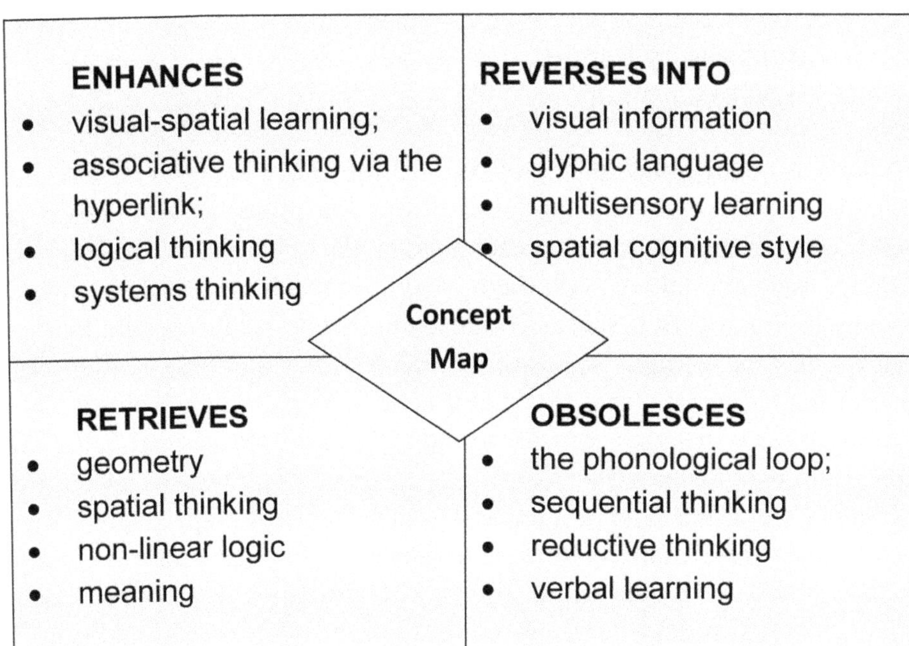

Or as a linear list, concept maps:

- Enhance
  - Visual-spatial learning;
  - Associative thinking via the hyperlink;
  - Logical thinking
  - Systems thinking
- Obsolesce
  - The phonological loop;
  - Sequential thinking
  - Reductive thinking
- Retrieve:
  - Geometry
  - Spatial thinking
  - Non-linear logic
  - Meaning
- Flip Into
  - Visual information
  - Glyphic language
  - Multisensory learning
  - SPATIAL learning

Notice as in any good conceptual model, the above analysis raises certain anomalies if we integrate the model with cognitive style theory. For example, according to the above tetrad, concept maps will flip or reverse into multisensory learning, which cognitive style theory would consider object learning. The anomaly in this regard is that object learning is opposed to spatial learning, so how can this incongruity be resolved? Another anomaly is that model predicts that concept maps, especially spatial maps, will obsolesce verbal learning. What will happen to science and technology studies and fields when science, along with numerical learning style, was founded on verbal learning?

In any case, it has been argued that spatial concept maps such as Visual Logic Maps are extremely powerful because although there are only a limited number of fundamental, dynamic cognitive processes represented by the glyphs, limitless iterations of these processes exist. In addition, the cognitive processes, represented by the glyphs and their corresponding maps may to be grounded in neural structures, generate relationships between bits of information, forming naturally occurring, often unique patterns of information. Cognitive processes visually drawn from the primitive seven glyphs in the construction of Visual Logic Maps are more congruent with underlying patterns of thinking than are the ideas and concepts manifested in linear written or spoken text and the verbal concept maps that interpret these tests. In a telephone conversation with David Hyerle, the inventor of Thinking Maps, Dr. Hyerle said that one way to visualize the reason certain concept maps are so effective in learning is to imagine each glyph and map as a chess piece with its own moves or functions. Although the number of pieces is limited, the glyphs and maps can be combined in limitless ways to form or analyze thoughts manifested as text and other media. These symbols of thought, the glyphs and their corresponding maps, like chess pieces, can express meaning ranging from the very basic and elementary to the extremely complex and sophisticated.

This book concludes with a concept map, in its simplest form a classification map.

In Figure 33 it is hypothesized that Lullian object maps evolved into Ramean verbal maps to Novak's concept maps, which were retrieved in the 20th century by Albert Upton and J.D. Novak, the inventor of the term concept map.

Figure 33. The evolution of concept mapping of concept mapping from the Middle Ages to today and from Object maps to Verbal Maps

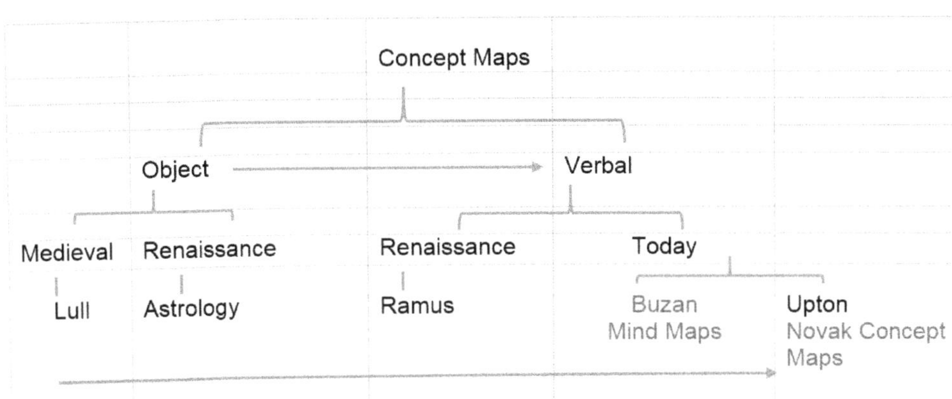

Figure 34. The evolution of concept maps fom verbal to spatial in modern times (Venn and Euler) to today (Visual Learning Maps (vLms)

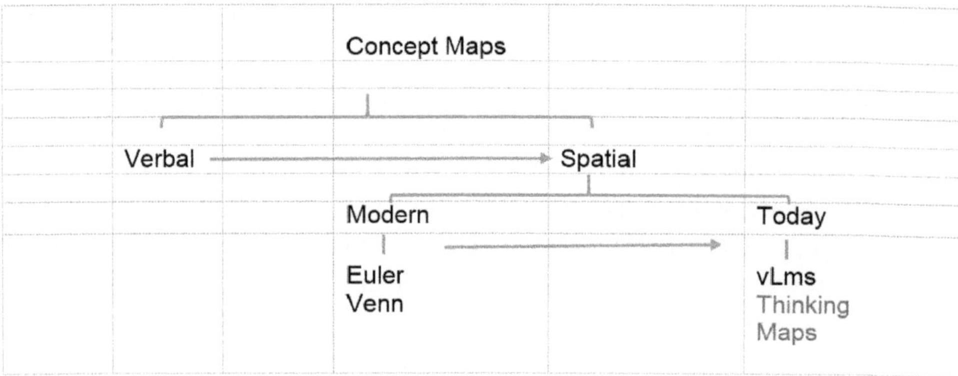

From the map analysis of Figure 34 it can be said that the very specific domain spatial maps of Euler and Venn evolved into modern general spatial maps exemplified by the Visual Logic Maps (vlms) presented and discussed in the concluding chapter.

REFERENCES

Bricken, W. (2005). *Taking nothing seriously*. Available at http://iconicmath.com/mypdfs/bl-serious.060105.pdf

Brown, G. S. (1972). *Laws of form*. The Julien Press Inc.

Budd, J. W. (2004). Mind Maps As Classroom Exercises. *The Journal of Economic Education*, *35*(1), 35–46. doi:10.3200/JECE.35.1.35-46

Buzan, T. (2006). *The ultimate book of mind maps: unlock your creativity, boost your memory, change your life*. HarperCollins, UK.

Buzan, T., & Buzan, B. (1996). *How to use radiant thinking to maximize your brain's untapped potential*. Plume Books.

Carnot, M. J., Feltovich, P., Hoffman, R. R., Feltovich, J., & Novak, J. D. (2003). *A summary of literature pertaining to the use of concept mapping techniques and technologies for education and performance support*. Pensacola, FL: Academic Press.

Centra, J. A., & Rock, D. (1971). College environments and student academic achievement. *American Educational Research Journal*, *8*(4), 623–634. doi:10.3102/00028312008004623

Chung, G. K. W. K., ONeil, H. F. Jr, & Herl, H. E. (1999). The use of computer-based collaborative knowledge mapping to measure team processes and team outcomes. *Computers in Human Behavior*, *15*(3), 463–493. doi:10.1016/S0747-5632(99)00032-1

Cohen, J. (1954). On the project of a universal character. *Mind*, *63*(249), 49–63. doi:10.1093/mind/LXIII.249.49

Conallen, J. (1999). Modeling Web application architectures with UML. *Communications of the ACM*, *42*(10), 63–70. doi:10.1145/317665.317677

Conklin, J. (2003). Dialog mapping: Reflections on an industrial strength case study. In *Visualizing argumentation* (pp. 117–136). Springer London. doi:10.1007/978-1-4471-0037-9_6

Davies, M. (2011). Concept mapping, mind mapping and argument mapping: What are the differences and do they matter? *Higher Education*, *62*(3), 279–301. doi:10.1007/s10734-010-9387-6

Embley, D. W. (1980, May). Programming with data frames for everyday data items. In *Proceedings of the May 19-22, 1980, national computer conference* (pp. 301-305). ACM. doi:10.1145/1500518.1500564

Eppler, M. J. (2006). A comparison between concept maps, mind maps, conceptual diagrams, and visual metaphors as complementary tools for knowledge construction and sharing. *Information Visualization*, *5*(3), 202-210.

Goodale, M. A., & Milner, A. D. (1992). Separate visual pathways for perception and action. *Trends in Neurosciences*, *15*(1), 20–25. doi:10.1016/0166-2236(92)90344-8 PMID:1374953

Gruber, T. R. (1993). A translation approach to portable ontology specifications. *Knowledge Acquisition*, *5*(2), 199–220. doi:10.1006/knac.1993.1008

Herl, H. E., ONeil, H. F. Jr, Chung, G. K. W. K., & Schacter, J. (1999). Reliability and validity of a computer-based knowledge mapping system to measure content understanding. *Computers in Human Behavior*, *15*(3), 315–333. doi:10.1016/S0747-5632(99)00026-6

Hung, J. L. (2012). Trends of e-learning research from 2000 to 2008: Use of text mining and bibliometrics. *British Journal of Educational Technology*, *43*(1), 5–16. doi:10.1111/j.1467-8535.2010.01144.x

Hwang, G. J., Yang, L. H., & Wang, S. Y. (2013). A concept map-embedded educational computer game for improving students learning performance in natural science courses. *Computers & Education*, *69*, 121–130. doi:10.1016/j.compedu.2013.07.008

Hyerle, D. (1996). *Visual Tools for Constructing Knowledge*. Alexandria, VA: Pitt Street.

Hyerle, D. (2014). Thinking Maps®: A Visual Language for Learning. In A. Okada, S. B. Shum, & T. Sherborne (Eds.), *Knowledge Cartography* (pp. 73–88). London: Springer. doi:10.1007/978-1-4471-6470-8_4

Hyerle, D. (2016). *David Hyerle, Design for Thinking*. Available at: http://dft.designsforthinking.com/?page_id=199

IHMC. (2016). *Joseph Novak*. Institute for Human and Machine Cognition. Available at https://www.ihmc.us/groups/jnovak/

Jones, B. D., Ruff, C., & Sunder, J. D. (2012). The Effects of Mind Mapping Activities on Students Motivation. *International Journal for the Scholarship of Teaching and Learning*, *6*(1), 1–21. doi:10.20429/ijsotl.2012.060105

Katifori, A., Halatsis, C., Lepouras, G., Vassilakis, C., & Giannopoulou, E. (2007). Ontology visualization methods - A Survey. *ACM Computing Surveys*, *39*(4), 10, es. doi:10.1145/1287620.1287621

Kawahara, M., & Matsuoka, K. (2013). Object-Spatial Imagery Types of Japanese College Students. *Psychology (Savannah, Ga.)*, *4*(03), 165–168.

Korzybski, A. (1958). *Science and sanity: An introduction to non-Aristotelian systems and general semantics*. Institute of GS.

Kozhevnikov, M., Blazhenkova, O., & Becker, M. (2010). Trade-off in object versus spatial visualization abilities: Restriction in the development of visual-processing resources. *Psychonomic Bulletin & Review*, *17*(1), 29–35. doi:10.3758/PBR.17.1.29 PMID:20081157

Kuhn, T. S. (2012). *The structure of scientific revolutions*. University of Chicago Press. doi:10.7208/chicago/9780226458144.001.0001

Lambiotte, J., Skaggs, L., & Dansereau, D. (1993). Learning from lectures: Effects of language vocabulary learning. *Journal of Experimental Education, 70*(1), 5–24.

Lemon, O. J. (1997). Logic and visual information. *Journal of Logic Language and Information, 6*(2), 213–216. doi:10.1023/A:1008278914765

McLuhan, M. (1994). *Understanding media: The extensions of man*. MIT Press.

McLuhan, M., & McLuhan, E. (1992). *Laws of media: The new science*. University of Toronto Press.

Novak, J. D., & Cañas, A. J. (2008). *The Theory Underlying Concept Maps and How to Construct and Use Them. Technical Report. Rev 01-2008*. Florida Institute for Human and Machine Cognition.

Ogiela, L., & Ogiela, M. R. (2012). *Advances in cognitive information systems* (Vol. 17). Springer Science & Business Media. doi:10.1007/978-3-642-25246-4

Okada, A., Shum, B. S., & Sherborne, T. (Eds.). (2008). Knowledge Cartography: Software tools and mapping techniques. London, UK: Springer.

Ong, W. J. (2005). *Ramus, method, and the decay of dialogue: From the art of discourse to the art of reason*. University of Chicago Press.

OText. (2016). Available at https://www.otexts.org/node/834

Sierhuis, M., & Shum, S. B. (2008). Human-agent knowledge cartography for e-science: NASA field trials at the Mars Desert Research Station. In *Knowledge Cartography* (pp. 287–305). Springer London. doi:10.1007/978-1-84800-149-7_14

Stapleton, G. (2005). A survey of reasoning systems based on Euler diagrams. *Electronic Notes in Theoretical Computer Science, 134*, 127–151. doi:10.1016/j.entcs.2005.02.022

Sternberg, R. J., & Powell, J. S. (1983). Comprehending verbal comprehension. *The American Psychologist, 38*(8), 878–893. doi:10.1037/0003-066X.38.8.878

Suthers, D., Connelly, J., Lesgold, A., Paolucci, M., Toth, E., Toth, J., & Weiner, A. (2001). Representational and advisory guidance for students learning scientific inquiry. *Smart machines in education: The coming revolution in educational technology*, 7-35..

Tang, M., Tran, T., Kim, H. J., Hund, A., & Knaus, K. (2012a). eLmaps: An Educational Software Application Based on Principles of Cognitive Linguistics and Cognitive Science. International Journal of Technology. *Knowledge in Society*, *8*(5).

Tang, M., Zhao, Y., & Costa, A. L. (2013, March). Efficient Learning Maps (eLms)©: A software application to learn and teach Visual Logic Maps (vLms)© with preliminary assessment results. In *Society for Information Technology & Teacher Education International Conference* (Vol. 2013, No. 1, pp. 95-106).

Tang, M. S., Hyerle, D., & Tran, T. (2012). A mathematical analysis of semantic maps, with theoretical and applied implications for blended learning software. *International Journal on E-Learning*, *11*(1), 95.

Tang, M. S., & Karunanithi, A. T. (2017In press). A media interpretation of the negative impact of cloud technologies. In L. Chao (Ed.), *Handbook of research on cloud-based STEM Education for improved learning outcomes* (pp. 505–422). IGI Global.

Upton, A. (1941). *Design for thinking: A first book in semantics*. Stanford University Press.

Upton, A., Farmer, A. D., & Samson, R. W. (1978). *Creative analysis*. Dutton.

Wang, H. Y., Huang, I., & Hwang, G. J. (2016). Effects of a Question Prompt-based Concept Mapping Approach on Students' Learning Achievements, Attitudes and 5C Competences in Project-based Computer Course Activities. *Journal of Educational Technology & Society*, *19*(3), 351–365.

Watson, M. K., Pelkey, J., Noyes, C. R., & Rodgers, M. O. (2016). Assessing Conceptual Knowledge Using Three Concept Map Scoring Methods. *The Journal of Engineering Education*, *105*(1), 118–146. doi:10.1002/jee.20111

Wheeler, L. A., & Collins, S. K. (2003). The influence of concept mapping on critical thinking in baccalaureate nursing students. *Journal of Professional Nursing*, *19*(6), 339–346. doi:10.1016/S8755-7223(03)00134-0 PMID:14689390

Wilson, R. (2008). *Lewis Carroll in Numberland: His Fantastical Mathematical Logical Life*. Allen Lane.

Wolfe, P. (2011). *Student Successes With Thinking Maps®: School-Based Research, Results, and Models for Achievement Using Visual Tools* (D. N. Hyerle & L. Alper, Eds.). Corwin Press.

Woolfolk, A. E., Winne, P. H., Perry, N. E., & Shapka, J. (2010). *Educational Psychology* (4th ed.). Toronto: Pearson Canada.

Yang, C. C., Hwang, G. J., Hung, C. M., & Tseng, S. S. (2013). An evaluation of the learning effectiveness of concept map-based science book reading via mobile devices. *Journal of Educational Technology & Society*, *16*(3), 167–178.

Related Readings

To continue IGI Global's long-standing tradition of advancing innovation through emerging research, please find below a compiled list of recommended IGI Global book chapters and journal articles in the areas of STEM education and science education. These related readings will provide additional information and guidance to further enrich your knowledge and assist you with your own research.

Akers, T., Nyarko, K., Gwynn, D., & Larkin, W. D. (2014). Visual Analytics for Students and Teachers (VAST) Model at a Minority Serving Institution: A Department of Homeland Security Project for Strengthening STEM Teaching and Learning. In C. Brown, K. Peters, & K. Nyarko (Eds.), *Cases on Research and Knowledge Discovery: Homeland Security Centers of Excellence* (pp. 152–169). Hershey, PA: IGI Global. doi:10.4018/978-1-4666-5946-9.ch007

Alkadi, I. (2016). Assessing Security with Regard to Cloud Applications in STEM Education. In L. Chao (Ed.), *Handbook of Research on Cloud-Based STEM Education for Improved Learning Outcomes* (pp. 260–276). Hershey, PA: IGI Global. doi:10.4018/978-1-4666-9924-3.ch017

Anohah, E. (2016). Pedagogy and Design of Online Learning Environment in Computer Science Education for High Schools. *International Journal of Online Pedagogy and Course Design*, *6*(3), 39–51. doi:10.4018/IJOPCD.2016070104

Anohah, E. (2017). Paradigm and Architecture of Computing Augmented Learning Management System for Computer Science Education. *International Journal of Online Pedagogy and Course Design*, *7*(2), 60–70. doi:10.4018/IJOPCD.2017040105

Anohah, E., & Suhonen, J.Solomon Sunday Oyelere. (2017). Trends of Mobile Learning in Computing Education from 2006 to 2014: A Systematic Review of Research Publications. *International Journal of Mobile and Blended Learning*, *9*(1), 16–33. doi:10.4018/IJMBL.2017010102

Arinze, B., Sylla, C., & Amobi, O. (2016). Cloud Computing for Teaching and Learning: Design Strategies. In L. Chao (Ed.), *Handbook of Research on Cloud-Based STEM Education for Improved Learning Outcomes* (pp. 159–171). Hershey, PA: IGI Global. doi:10.4018/978-1-4666-9924-3.ch011

Ault, M., Craig-Hare, J., Ellis, J. D., Bulgren, J., Kretschmer, I., & Frey, B. B. (2016). Using Reason Racer to Support Argumentation in Middle School Science Instruction. In M. Urban & D. Falvo (Eds.), *Improving K-12 STEM Education Outcomes through Technological Integration* (pp. 399–431). Hershey, PA: IGI Global. doi:10.4018/978-1-4666-9616-7.ch018

Back, S. M., Greenhalgh-Spencer, H., & Frias, K. M. (2016). The Application of Transdisciplinary Theory and Practice to STEM Education. In Y. Rosen, S. Ferrara, & M. Mosharraf (Eds.), *Handbook of Research on Technology Tools for Real-World Skill Development* (pp. 42–67). Hershey, PA: IGI Global. doi:10.4018/978-1-4666-9441-5.ch003

Barker, B. S., Nugent, G., Grandgenett, N., Keshwani, J., Nelson, C. A., & Leduc-Mills, B. (2016). Developing an Elementary Engineering Education Program through Problem-Based Wearable Technologies Activities. In J. Holland (Ed.), *Wearable Technology and Mobile Innovations for Next-Generation Education* (pp. 269–294). Hershey, PA: IGI Global. doi:10.4018/978-1-5225-0069-8.ch014

Bennett, A., Eglash, R., Lachney, M., & Babbitt, W. (2016). Design Agency: Diversifying Computer Science at the Intersections of Creativity and Culture. In M. Raisinghani (Ed.), *Revolutionizing Education through Web-Based Instruction* (pp. 35–56). Hershey, PA: IGI Global. doi:10.4018/978-1-4666-9932-8.ch003

Berdanier, C. G., Zephirin, T., Cox, M. F., & Black, S. M. (2015). Teaching MSE Students to Teach: A Design-Based Research Model for Introducing Professional Skills into the Technical Curriculum. In H. Lim (Ed.), *Handbook of Research on Recent Developments in Materials Science and Corrosion Engineering Education* (pp. 369–396). Hershey, PA: IGI Global. doi:10.4018/978-1-4666-8183-5.ch019

Bernhardt, S. (2014). Structural Context and Influencing Factors. In *Women in IT in the New Social Era: A Critical Evidence-Based Review of Gender Inequality and the Potential for Change* (pp. 127–157). Hershey, PA: IGI Global. doi:10.4018/978-1-4666-5860-8.ch006

Beziat, T. L., McCombs, K. M., Burks, B. A., & Byrom, J. (2016). Are Female STEM Majors Academic Risk Takers? In U. Thomas & J. Drake (Eds.), *Critical Research on Sexism and Racism in STEM Fields* (pp. 144–152). Hershey, PA: IGI Global. doi:10.4018/978-1-5225-0174-9.ch008

Brewley, D. N., Boindala, P. S., & Sinclair, J. L. (2015). Ideation to Execution: Flipping an Undergraduate Pre-Calculus Course to Create Significant Learning Experiences. In A. Scheg (Ed.), *Implementation and Critical Assessment of the Flipped Classroom Experience* (pp. 25–40). Hershey, PA: IGI Global. doi:10.4018/978-1-4666-7464-6.ch002

Brown, C. A. (2016). Using Digital Resources to Support STEM Education. In V. Wang (Ed.), *Handbook of Research on Learning Outcomes and Opportunities in the Digital Age* (pp. 127–151). Hershey, PA: IGI Global. doi:10.4018/978-1-4666-9577-1.ch006

Buck, G. A., Beeman-Cadwallader, N., & Trauth-Nare, A. (2016). Examining the Levels of Reasoning Used by Urban Elementary Black Girls Engaging in Technology-Enhanced Inquiry. In M. Urban & D. Falvo (Eds.), *Improving K-12 STEM Education Outcomes through Technological Integration* (pp. 86–107). Hershey, PA: IGI Global. doi:10.4018/978-1-4666-9616-7.ch005

Cai, T. (2015). From Learning Management System to Internet-Based Research in Mathematics Education. In A. Scheg (Ed.), *Critical Examinations of Distance Education Transformation across Disciplines* (pp. 62–85). Hershey, PA: IGI Global. doi:10.4018/978-1-4666-6555-2.ch004

Cartledge, D. (2014). Re–Valuing Constructed Knowledge: An Agenda for STE(A)M. In T. Fitzgerald (Ed.), *Advancing Knowledge in Higher Education: Universities in Turbulent Times* (pp. 170–186). Hershey, PA: IGI Global. doi:10.4018/978-1-4666-6202-5.ch011

Chang, S., von Imhoff, M., & Ustrup, R. I. (2016). Engineering and Information Technology: Challenges and Opportunities for Exchange Studies. In D. Velliaris & D. Coleman-George (Eds.), *Handbook of Research on Study Abroad Programs and Outbound Mobility* (pp. 300–324). Hershey, PA: IGI Global. doi:10.4018/978-1-5225-0169-5.ch012

Chapman, D. L., & Wang, S. (2015). Multimedia Instructional Tools and Student Learning in a Computer Applications Course. *International Journal of Information and Communication Technology Education, 11*(2), 57–67. doi:10.4018/ijicte.2015040105

Charles, D. R., & Wray, F. (2015). The English Science Cities: A New Phase in Science-based Urban Strategy. *International Journal of Knowledge-Based Organizations, 5*(1), 46–61. doi:10.4018/ijkbo.2015010104

Chen, Y. (2016). Security Issues Related to Cloud Applications in STEM Education. In L. Chao (Ed.), *Handbook of Research on Cloud-Based STEM Education for Improved Learning Outcomes* (pp. 277–289). Hershey, PA: IGI Global. doi:10.4018/978-1-4666-9924-3.ch018

Choi, H., & Feng, J. (2017). Using Video Games to Improve Spatial Skills. In R. Zheng & M. Gardner (Eds.), *Handbook of Research on Serious Games for Educational Applications* (pp. 93–114). Hershey, PA: IGI Global. doi:10.4018/978-1-5225-0513-6.ch005

Choi, J., An, S., & Lee, Y. (2015). A Case Study of Distance Education for Informatics Gifted Students. In L. Lennex & K. Nettleton (Eds.), *Cases on Instructional Technology in Gifted and Talented Education* (pp. 358–380). Hershey, PA: IGI Global. doi:10.4018/978-1-4666-6489-0.ch017

Coleman, N., Farina, D. M., & Rabinovich, L. (2017). Common Denominators to Learner-Centered Success: Undergraduate STEM, Graduate Teacher Education, and an Educational Technology Doctoral Program. In J. Keengwe & G. Onchwari (Eds.), *Handbook of Research on Learner-Centered Pedagogy in Teacher Education and Professional Development* (pp. 23–44). Hershey, PA: IGI Global. doi:10.4018/978-1-5225-0892-2.ch002

Coyle, V. C., Newman, D. L., & Connor, K. A. (2016). Innovative Instruction in STEM Education: The Role of Student Feedback in the Development of a Flipped Classroom. In J. Keengwe & G. Onchwari (Eds.), *Handbook of Research on Active Learning and the Flipped Classroom Model in the Digital Age* (pp. 309–332). Hershey, PA: IGI Global. doi:10.4018/978-1-4666-9680-8.ch016

Dass, P. M., & Spagnolo, J. T. (2016). STEM Learning in Middle Grades by Technology-Mediated Integration of Science and Mathematics: Results of Project SMILE. In M. Urban & D. Falvo (Eds.), *Improving K-12 STEM Education Outcomes through Technological Integration* (pp. 187–205). Hershey, PA: IGI Global. doi:10.4018/978-1-4666-9616-7.ch009

Dawson, M., Al Saeed, I., Wright, J., & Onyegbula, F. (2014). Open Source Software to Enhance the STEM Learning Environment. In V. Wang (Ed.), *Handbook of Research on Education and Technology in a Changing Society* (pp. 569–580). Hershey, PA: IGI Global. doi:10.4018/978-1-4666-6046-5.ch042

De la Garza, R. L., Zavala, G., & Gómez Galindo, A. A. (2017). Argumentation Schema to Analyze High School Students' Scientific Reasoning. In M. Ramírez-Montoya (Ed.), *Handbook of Research on Driving STEM Learning With Educational Technologies* (pp. 297–314). Hershey, PA: IGI Global. doi:10.4018/978-1-5225-2026-9.ch015

DeCoito, I., & Richardson, T. (2016). Using Technology to Enhance Science Literacy, Mathematics Literacy, or Technology Literacy: Focusing on Integrated STEM Concepts in a Digital Game. In M. Urban & D. Falvo (Eds.), *Improving K-12 STEM Education Outcomes through Technological Integration* (pp. 1–22). Hershey, PA: IGI Global. doi:10.4018/978-1-4666-9616-7.ch001

Delgado-Cepeda, F. J. (2016). Upgrading a Numerical Methods Course into New Mobile Technologies for Mathematical Education: An Approach Based on Flexibility and Skill Development. In D. Parsons (Ed.), *Mobile and Blended Learning Innovations for Improved Learning Outcomes* (pp. 201–222). Hershey, PA: IGI Global. doi:10.4018/978-1-5225-0359-0.ch011

Denner, J., Martinez, J., & Thiry, H. (2017). Strategies for Engaging Hispanic/Latino Youth in the US in Computer Science. In Y. Rankin & J. Thomas (Eds.), *Moving Students of Color from Consumers to Producers of Technology* (pp. 24–48). Hershey, PA: IGI Global. doi:10.4018/978-1-5225-2005-4.ch002

Deyoe, M. M., Newman, D. L., & Lamendola, J. M. (2015). Mobile Technology in Higher Education: Patterns of Replication and Transferability. In J. Keengwe & M. Maxfield (Eds.), *Advancing Higher Education with Mobile Learning Technologies: Cases, Trends, and Inquiry-Based Methods* (pp. 128–153). Hershey, PA: IGI Global. doi:10.4018/978-1-4666-6284-1.ch007

Dietz, G., Hessedence, J., Long, T., & Muga, H. E. (2017). Empowering Women in STEM: Embedding STEM in K-12 Education. In M. Gray & K. Thomas (Eds.), *Strategies for Increasing Diversity in Engineering Majors and Careers* (pp. 61–87). Hershey, PA: IGI Global. doi:10.4018/978-1-5225-2212-6.ch004

Dollahon, C. V. (2017). Using STEAM in Marine Science: Incorporating Graphic Design Into an Existing STEM Lesson. In J. Bazler & M. Van Sickle (Eds.), *Cases on STEAM Education in Practice* (pp. 292–317). Hershey, PA: IGI Global. doi:10.4018/978-1-5225-2334-5.ch015

Dongale, T. D., Patil, S. S., & Kamat, R. K. (2015). Learning by Simulations: A New and Effective Pedagogical Approach for Science, Engineering and Technology Students in a Traditional Setting. *International Journal of Quality Assurance in Engineering and Technology Education*, *4*(2), 13–25. doi:10.4018/IJQAETE.2015040102

Dragnic-Cindric, D., Barrow, E., & Anderson, J. L. (2017). Opportunity to Start Strong: Integration of Technology in Science Lessons in the Early Elementary Grades. In J. Keengwe & P. Bull (Eds.), *Handbook of Research on Transformative Digital Content and Learning Technologies* (pp. 154–170). Hershey, PA: IGI Global. doi:10.4018/978-1-5225-2000-9.ch009

Duran, A., & Lopez, D. (2015). Women from Diverse Backgrounds in the Science, Technology, Engineering, and Math (STEM) Professions: Retention and Career Development. In C. Hughes (Ed.), *Impact of Diversity on Organization and Career Development* (pp. 214–251). Hershey, PA: IGI Global. doi:10.4018/978-1-4666-7324-3.ch009

Dzan, W., Tsai, H., Lou, S., & Shih, R. (2015). Satisfaction Analysis of Experiential Learning-Based Popular Science Education. *International Journal of Distance Education Technologies*, *13*(2), 93–109. doi:10.4018/IJDET.2015040106

Eglash, R., Babbitt, W., Bennett, A., Bennett, K., Callahan, B., Davis, J., & Tully, K. et al. (2017). Culturally Situated Design Tools: Generative Justice as a Foundation for STEM Diversity. In Y. Rankin & J. Thomas (Eds.), *Moving Students of Color from Consumers to Producers of Technology* (pp. 132–151). Hershey, PA: IGI Global. doi:10.4018/978-1-5225-2005-4.ch007

Eguchi, A. (2015). Educational Robotics as a Learning Tool for Promoting Rich Environments for Active Learning (REALs). In J. Keengwe (Ed.), *Handbook of Research on Educational Technology Integration and Active Learning* (pp. 19–47). Hershey, PA: IGI Global. doi:10.4018/978-1-4666-8363-1.ch002

El-Farargy, N. (2015). Making Sense of Science: A Review in Scottish Further Education. In E. de Silva (Ed.), *Cases on Research-Based Teaching Methods in Science Education* (pp. 1–29). Hershey, PA: IGI Global. doi:10.4018/978-1-4666-6375-6.ch001

Erickson, J. J. (2015). To Play or to Learn?: A Review of Game-Based Math Learning for Motivation and Cognition. *International Journal of Cyber Behavior, Psychology and Learning*, *5*(1), 56–74. doi:10.4018/ijcbpl.2015010105

Estevez, C. (2016). Addressing Transport Layer Issues in Cloud Computing: A STEM Perspective. In L. Chao (Ed.), *Handbook of Research on Cloud-Based STEM Education for Improved Learning Outcomes* (pp. 79–93). Hershey, PA: IGI Global. doi:10.4018/978-1-4666-9924-3.ch006

Fakayode, S. O., Snipes, V., Kanipes, M. I., Mohammed, A. K., & Wilson, Z. S. (2016). Use of Innovative Pedagogies and Creative Partnership Strategies to Promote Undergraduate STEM Education at an HBCU. In C. Prince & R. Ford (Eds.), *Setting a New Agenda for Student Engagement and Retention in Historically Black Colleges and Universities* (pp. 119–151). Hershey, PA: IGI Global. doi:10.4018/978-1-5225-0308-8.ch008

Fakomogbon, M. A., Adebayo, R. F., Adegbija, M. V., Shittu, A. T., & Oyelekan, O. S. (2014). Subject Specialization and Science Teachers Perception of Information and Communication Technology for Instruction in Kwara State. *International Journal of Information and Communication Technology Education*, *10*(1), 25–34. doi:10.4018/ijicte.2014010103

Farland-Smith, D., & Ledger, T. (2016). Understanding How Images and Attitudes toward Scientists and Science Contribute to Science Identities: Investigating How Images Drawn by Elementary, Middle, and High School Students Reflect Their Attitudes. In E. Railean, G. Walker, A. Elçi, & L. Jackson (Eds.), *Handbook of Research on Applied Learning Theory and Design in Modern Education* (pp. 498–518). Hershey, PA: IGI Global. doi:10.4018/978-1-4666-9634-1.ch024

Feldhaus, C. R., Little, J., & Sorge, B. (2015). Conflict Resolution and Ethical Decision-Making for Engineering Professionals in Global Organizations. In S. Sundar Sethy (Ed.), *Contemporary Ethical Issues in Engineering* (pp. 204–227). Hershey, PA: IGI Global. doi:10.4018/978-1-4666-8130-9.ch015

Fleenor, H. G., & Hodhod, R. (2016). Assessment of Learning and Technology: Computer Science Education. In V. Wang (Ed.), *Handbook of Research on Learning Outcomes and Opportunities in the Digital Age* (pp. 51–78). Hershey, PA: IGI Global. doi:10.4018/978-1-4666-9577-1.ch003

Fletcher, E. C. Jr, & Djajalaksana, Y. (2014). Predictors of Instructional Strategy Use of Faculty in Career and Technical Education Programs: Signature Pedagogies of the Field. *International Journal of Adult Vocational Education and Technology*, *5*(3), 45–65. doi:10.4018/ijavet.2014070104

Flournoy, B. (2016). Black Women and Science in Higher Education: Not Much Has Changed – Still the Struggle. In U. Thomas & J. Drake (Eds.), *Critical Research on Sexism and Racism in STEM Fields* (pp. 170–180). Hershey, PA: IGI Global. doi:10.4018/978-1-5225-0174-9.ch010

Gabriel, B. F., Valente, R. A., Dias-de-Oliveira, J., Neto, V. F., & Andrade-Campos, A. (2017). Methodologies for Engineering Learning and Teaching (MELT): An Overview of Engineering Education in Europe and a Novel Concept for Young Students. In L. Carvalho (Ed.), *Handbook of Research on Entrepreneurial Development and Innovation Within Smart Cities* (pp. 363–391). Hershey, PA: IGI Global. doi:10.4018/978-1-5225-1978-2.ch016

Galeshi, R. (2017). Engineering Education for All: Increasing Access to Engineering Education for Men and Women across the World through Distance Learning. *International Journal of Online Pedagogy and Course Design*, *7*(2), 35–47. doi:10.4018/IJOPCD.2017040103

Garner, N., Lischke, M. D., Siol, A., & Eilks, I. (2014). Learning about Sustainability in a Non-Formal Laboratory Context for Secondary Level Students: A Module on Climate Change, the Ozone Hole, and Summer Smog. In K. Thomas & H. Muga (Eds.), *Handbook of Research on Pedagogical Innovations for Sustainable Development* (pp. 229–244). Hershey, PA: IGI Global. doi:10.4018/978-1-4666-5856-1.ch012

Gatling, A. P. (2016). Technology's Role in Supporting Elementary Preservice Teachers as They Teach: An Urban STEM Afterschool Enrichment Program. In M. Urban & D. Falvo (Eds.), *Improving K-12 STEM Education Outcomes through Technological Integration* (pp. 362–379). Hershey, PA: IGI Global. doi:10.4018/978-1-4666-9616-7.ch016

Gomes, A. D., Mendes, A. J., & Marcelino, M. J. (2015). Computer Science Education Research: An Overview and Some Proposals. In R. Queirós (Ed.), *Innovative Teaching Strategies and New Learning Paradigms in Computer Programming* (pp. 1–29). Hershey, PA: IGI Global. doi:10.4018/978-1-4666-7304-5.ch001

Gorman, J., & Heinze-Fry, J. (2014). Conceptual Mapping Facilitates Coherence and Critical Thinking in the Science Education System. In L. Shedletsky & J. Beaudry (Eds.), *Cases on Teaching Critical Thinking through Visual Representation Strategies* (pp. 296–334). Hershey, PA: IGI Global. doi:10.4018/978-1-4666-5816-5.ch012

Günel, M., Özer-Keskin, M., & Keskin-Samancı, N. (2017). A Paradigm Shift for Teachers' Professional Development Structure in Turkey: Moving from Instruction to Learning. In *Educational Leadership and Administration: Concepts, Methodologies, Tools, and Applications* (pp. 282–302). Hershey, PA: IGI Global. doi:10.4018/978-1-5225-1624-8.ch016

Haswell, M. M. (2016). Barriers to Success: Disadvantages of Gender Normative Language to Women in STEM. In U. Thomas & J. Drake (Eds.), *Critical Research on Sexism and Racism in STEM Fields* (pp. 49–66). Hershey, PA: IGI Global. doi:10.4018/978-1-5225-0174-9.ch003

Havard, B., East, M. L., Prayaga, L., & Whiteside, A. (2016). Adaptable Learning Theory Framework for Technology-Enhanced Learning. In E. Railean, G. Walker, A. Elçi, & L. Jackson (Eds.), *Handbook of Research on Applied Learning Theory and Design in Modern Education* (pp. 632–654). Hershey, PA: IGI Global. doi:10.4018/978-1-4666-9634-1.ch031

Herr, N., Tippens, M., Rivas, M., Vandergon, V. O., d'Alessio, M., & Reveles, J. (2016). Cloud-Based Continuous Formative Assessment (CFA). In L. Chao (Ed.), *Handbook of Research on Cloud-Based STEM Education for Improved Learning Outcomes* (pp. 311–342). Hershey, PA: IGI Global. doi:10.4018/978-1-4666-9924-3.ch020

Hoban, G., Nielsen, W., & Hyland, C. (2016). Blended Media: Student-Generated Mash-ups to Promote Engagement with Science Content. *International Journal of Mobile and Blended Learning*, *8*(3), 35–48. doi:10.4018/IJMBL.2016070103

Hoge, B. (2014). GBL as PBL: Guidelines for Game-Based Learning in the Classroom and Informal Science Centers. In Z. Yang, H. Yang, D. Wu, & S. Liu (Eds.), *Transforming K-12 Classrooms with Digital Technology* (pp. 58–82). Hershey, PA: IGI Global. doi:10.4018/978-1-4666-4538-7.ch004

Jablon, P. C. (2017). Theater as the STEAM Engine for Engaging Those Previously Disengaged. In J. Bazler & M. Van Sickle (Eds.), *Cases on STEAM Education in Practice* (pp. 118–153). Hershey, PA: IGI Global. doi:10.4018/978-1-5225-2334-5.ch006

Jiang, E. P. (2014). A Comparative Study on Undergraduate Computer Science Education between China and the United States. In V. Wang (Ed.), *International Education and the Next-Generation Workforce: Competition in the Global Economy* (pp. 208–223). Hershey, PA: IGI Global. doi:10.4018/978-1-4666-4498-4.ch012

Johnson, M. L., & Spear, K. (2017). Developing a Learning Community of Engineers Through an Honors First-Year Seminar. In M. Gray, & K. Thomas (Eds.), *Strategies for Increasing Diversity in Engineering Majors and Careers* (pp. 88-105). Hershey, PA: IGI Global. doi:10.4018/978-1-5225-2212-6.ch005

Juanes, J. A., & Ruisoto, P. (2014). Technological Advances and Teaching Innovation Applied to Health Science Education. *Journal of Information Technology Research*, *7*(2), 1–6. doi:10.4018/jitr.2014040101

Kalaian, S. A., & Kasim, R. M. (2015). Small-Group vs. Competitive Learning in Computer Science Classrooms: A Meta-Analytic Review. In R. Queirós (Ed.), *Innovative Teaching Strategies and New Learning Paradigms in Computer Programming* (pp. 46–64). Hershey, PA: IGI Global. doi:10.4018/978-1-4666-7304-5.ch003

Karlsson, G. (2015). Students Joint Reasoning about Gas Solubility in Water in Modified Versions of a Virtual Laboratory. *International Journal of Online Pedagogy and Course Design*, *5*(4), 67–77. doi:10.4018/IJOPCD.2015100105

Kennedy, B. F., & McLaughlin, P. M. (2017). Growing Global STEM Learning in Higher Education. In L. Leavitt, S. Wisdom, & K. Leavitt (Eds.), *Cultural Awareness and Competency Development in Higher Education* (pp. 132–147). Hershey, PA: IGI Global. doi:10.4018/978-1-5225-2145-7.ch008

Kleanthous, I., & Meletiou-Mavrotheris, M. (2016). Early Statistical Reasoning: An Exploratory Study of Primary School Students Use of a Dynamic Statistics Software Package for Analyzing and Interpreting Data. *International Journal of Information Communication Technologies and Human Development*, *8*(1), 26–41. doi:10.4018/IJICTHD.2016010102

Komura, K. (2017). Professional Masters' STEM Graduate Education Programs to Develop a Business-Savvy Workforce. In B. Narasimharao, E. Wright, S. Prasad, & M. Joshi (Eds.), *Handbook of Research on Science Education and University Outreach as a Tool for Regional Development* (pp. 35–49). Hershey, PA: IGI Global. doi:10.4018/978-1-5225-1880-8.ch004

Kumar, V., & Sharma, D. (2017). Cloud Computing as a Catalyst in STEM Education. *International Journal of Information and Communication Technology Education, 13*(2), 38–51. doi:10.4018/IJICTE.2017040104

Kumtepe, A. T., & Genc-Kumtepe, E. (2014). STEM in Early Childhood Education: We Talk the Talk, But Do We Walk the Walk? In Z. Yang, H. Yang, D. Wu, & S. Liu (Eds.), *Transforming K-12 Classrooms with Digital Technology* (pp. 140–163). Hershey, PA: IGI Global. doi:10.4018/978-1-4666-4538-7.ch008

Kvasny, L., Trauth, E. M., & Joshi, K. D. (2016). The Role of HBCUs in Preparing African American Males for Careers in Information Technology. In C. Prince & R. Ford (Eds.), *Setting a New Agenda for Student Engagement and Retention in Historically Black Colleges and Universities* (pp. 234–250). Hershey, PA: IGI Global. doi:10.4018/978-1-5225-0308-8.ch013

Laffey, J. M., Sadler, T. D., Goggins, S. P., Griffin, J., & Babiuch, R. N. (2016). Mission HydroSci: Distance Learning through Game-Based 3D Virtual Learning Environments. In D. Russell & J. Laffey (Eds.), *Handbook of Research on Gaming Trends in P-12 Education* (pp. 421–441). Hershey, PA: IGI Global. doi:10.4018/978-1-4666-9629-7.ch020

Lang, C., Craig, A., & Egan, M. (2016). The Importance of Outreach Programs to Unblock the Pipeline and Broaden Diversity in ICT Education. *International Journal of Information and Communication Technology Education, 12*(1), 38–49. doi:10.4018/IJICTE.2016010104

Leal, A. V., & Ferreira, D. J. (2016). Learning Programming Patterns Using Games. *International Journal of Information and Communication Technology Education, 12*(2), 23–34. doi:10.4018/IJICTE.2016040103

Lee, Y., Dunbar, N., Kornelson, K., Wilson, S. N., Ralston, R., Savic, M., & Elizondo, J. et al. (2016). Digital Game based Learning for Undergraduate Calculus Education: Immersion, Calculation, and Conceptual Understanding. *International Journal of Gaming and Computer-Mediated Simulations, 8*(1), 13–27. doi:10.4018/IJGCMS.2016010102

Lin, Y., Shen, M., & Liu, C. (2014). A Research of Employing Cognitive Load Theory in Science Education via Web-Pages. *International Journal of Online Pedagogy and Course Design, 4*(2), 19–34. doi:10.4018/ijopcd.2014040102

Liu, X., Xu, C., & Ouyang, B. (2015). Plagiarism Detection Algorithm for Source Code in Computer Science Education. *International Journal of Distance Education Technologies, 13*(4), 29–39. doi:10.4018/IJDET.2015100102

Lopez, C. (2015). Community Colleges and the Global Workforce: The Role and Impact of Community Colleges on the Pursuit of STEM Education and the Workforce. In S. Jones & D. Smith (Eds.), *Examining the Impact of Community Colleges on the Global Workforce* (pp. 209–228). Hershey, PA: IGI Global. doi:10.4018/978-1-4666-8481-2.ch011

Lou, S., Tsai, H., Tseng, K., & Shih, R. (2014). Effects of Implementing STEM-I Project-Based Learning Activities for Female High School Students. *International Journal of Distance Education Technologies, 12*(1), 52–73. doi:10.4018/ijdet.2014010104

Ma, J., Chiu, D. K., & Tang, J. K. (2016). Exploring the Use of Social Media to Advance K12 Science Education. *International Journal of Systems and Service-Oriented Engineering, 6*(4), 47–59. doi:10.4018/IJSSOE.2016100104

Mai, M. Y., Halim, L., Yaseen, R. M., & Meerah, T. S. (2014). Science, Technology, and Society (STS) Issues: A Survey of Secondary Students in Sana'a, Yemen. In M. Khosrow-Pour (Ed.), *Educational Technology Use and Design for Improved Learning Opportunities* (pp. 203–212). Hershey, PA: IGI Global. doi:10.4018/978-1-4666-6102-8.ch011

Marrero, M. E., Gunning, A. M., & Woodruff, K. (2016). Using Authentic Earth Data in the K-12 Classroom. In M. Urban & D. Falvo (Eds.), *Improving K-12 STEM Education Outcomes through Technological Integration* (pp. 281–309). Hershey, PA: IGI Global. doi:10.4018/978-1-4666-9616-7.ch013

Marshall, J. A., Erickson, T., & Sivam, K. (2015). Preservice Teachers Exploring the Nature of Science in Simulated Worlds. *International Journal of Gaming and Computer-Mediated Simulations*, *7*(2), 24–45. doi:10.4018/IJGCMS.2015040102

Martin, S. F., Green, A., & Dean, M. (2016). African American Women in STEM Education: The Cycle of Microaggressions from P-12 Classrooms to Higher Education and Back. In U. Thomas & J. Drake (Eds.), *Critical Research on Sexism and Racism in STEM Fields* (pp. 135–143). Hershey, PA: IGI Global. doi:10.4018/978-1-5225-0174-9.ch007

Martinez-Garza, M. M. (2015). Examining Epistemic Practices of the Community of Players of Dwarf Fortress: "For!! SCIENCE!! *International Journal of Gaming and Computer-Mediated Simulations*, *7*(2), 46–67. doi:10.4018/ijgcms.2015040103

Mazuritskiy, M., Safontsev, S., Konoplev, B., & Boldyreva, A. (2014). Remote Access to Scientific Laboratory Equipment and Competency-Based Approach to Science and Technology Education. *International Journal of Information and Communication Technology Education*, *10*(3), 19–33. doi:10.4018/ijicte.2014070102

Moallem, M., Morge, S. P., Narayan, S., & Tagliarini, G. A. (2016). The Power of Computational Modeling and Simulation for Learning STEM Content in Middle and High Schools. In M. Urban & D. Falvo (Eds.), *Improving K-12 STEM Education Outcomes through Technological Integration* (pp. 135–171). Hershey, PA: IGI Global. doi:10.4018/978-1-4666-9616-7.ch007

Mpofu, C. (2016). STEM Fields and Ethnic Women in New Zealand: Issues of Sexism and Racism. In U. Thomas & J. Drake (Eds.), *Critical Research on Sexism and Racism in STEM Fields* (pp. 181–207). Hershey, PA: IGI Global. doi:10.4018/978-1-5225-0174-9.ch011

Mtebe, J. S., & Kissaka, M. M. (2016). Enhancing the Quality of Computer Science Education with MOOCs in Sub-Saharan Africa. In J. Keengwe & G. Onchwari (Eds.), *Handbook of Research on Active Learning and the Flipped Classroom Model in the Digital Age* (pp. 366–377). Hershey, PA: IGI Global. doi:10.4018/978-1-4666-9680-8.ch019

Mulvenon, S. W., & Wang, V. C. (2015). Graduate Level Research Methods and Statistics Courses: The Perspective of an Instructor. *International Journal of Adult Vocational Education and Technology*, *6*(4), 1–11. doi:10.4018/IJAVET.2015100101

Mungai, J. N. (2015). Quality Preparation of Mathematics and Science Teachers to Integrate ICT: Lessons from Learner-Centered Teacher Professional Development Approach. In F. Nafukho & B. Irby (Eds.), *Handbook of Research on Innovative Technology Integration in Higher Education* (pp. 29–53). Hershey, PA: IGI Global. doi:10.4018/978-1-4666-8170-5.ch002

Neri de Souza, F. (2016). Science Education with and through ICT: Curriculum Design and Questioning to Promote Active Learning. In D. Fonseca & E. Redondo (Eds.), *Handbook of Research on Applied E-Learning in Engineering and Architecture Education* (pp. 1–14). Hershey, PA: IGI Global. doi:10.4018/978-1-4666-8803-2.ch001

Newman, D. L., Deyoe, M. M., Connor, K. A., & Lamendola, J. M. (2014). Flipping STEM Learning: Impact on Students' Process of Learning and Faculty Instructional Activities. In J. Keengwe, G. Onchwari, & J. Oigara (Eds.), *Promoting Active Learning through the Flipped Classroom Model* (pp. 113–131). Hershey, PA: IGI Global. doi:10.4018/978-1-4666-4987-3.ch006

Newman, D. L., Lamendola, J. M., Deyoe, M. M., & Connor, K. A. (2015). Active Learning, Mentoring, and Mobile Technology: Meeting Needs across Levels in One Place. In J. Keengwe (Ed.), *Promoting Active Learning through the Integration of Mobile and Ubiquitous Technologies* (pp. 116–134). Hershey, PA: IGI Global. doi:10.4018/978-1-4666-6343-5.ch007

Ojeda-Castro, A. M., Murray-Finley, P., & Sánchez-Villafañe, J. (2017). Learning Management System Use to Increase Mathematics Knowledge and Skills in Puerto Rico. *International Journal of Technology and Human Interaction*, *13*(2), 89–100. doi:10.4018/IJTHI.2017040106

Ojokoh, B. A., Adeola, O. S., Isinkaye, F. O., & Abraham, C. (2014). Career Choices in Information and Communication Technology among South Western Nigerian Women. *Journal of Global Information Management*, *22*(2), 48–77. doi:10.4018/jgim.2014040104

Osisioma, I. U. (2017). Africanizing Science Education: Engaging Students in Context-Based Science Instruction. In R. Gordon, T. Akutsu, J. McDermott, & J. Lalas (Eds.), *Challenges Associated with Cross-Cultural and At-Risk Student Engagement* (pp. 120–147). Hershey, PA: IGI Global. doi:10.4018/978-1-5225-1894-5.ch007

Oyelekan, O. S., Akinpelu, G. A., & Daramola, F. O. (2015). Science Students Use of the Internet for Learning in Higher Institutions in Osun State, Nigeria. *International Journal of Information and Communication Technology Education*, *11*(4), 67–82. doi:10.4018/IJICTE.2015100105

Peters, K. A., Wright Brown, C., & Nyarko, K. (2014). Conclusion: Perspectives on Effective Collaborative STEM Research Experiences Linked to DHS Centers of Excellence (COE). In C. Brown, K. Peters, & K. Nyarko (Eds.), *Cases on Research and Knowledge Discovery: Homeland Security Centers of Excellence* (pp. 268–292). Hershey, PA: IGI Global. doi:10.4018/978-1-4666-5946-9.ch012

Popov, O. (2015). Outdoor Science in Teacher Education. In T. Hansson (Ed.), *Contemporary Approaches to Activity Theory: Interdisciplinary Perspectives on Human Behavior* (pp. 128–142). Hershey, PA: IGI Global. doi:10.4018/978-1-4666-6603-0.ch008

Pugh, V. M. (2014). Lessons from the STEM Sector. In J. Prescott & J. McGurren (Eds.), *Gender Considerations and Influence in the Digital Media and Gaming Industry* (pp. 175–185). Hershey, PA: IGI Global. doi:10.4018/978-1-4666-6142-4.ch009

Qian, Y. (2015). 3D Multi-User Virtual Environments in Science Education: Potential and Challenges. In T. Heafner, R. Hartshorne, & T. Petty (Eds.), *Exploring the Effectiveness of Online Education in K-12 Environments* (pp. 302–324). Hershey, PA: IGI Global. doi:10.4018/978-1-4666-6383-1.ch015

Rappaport, J. M., Richter, S. B., & Kennedy, D. T. (2016). A Strategic Perspective on Using Symbolic Transformation in STEM Education: Robotics and Automation. *International Journal of Strategic Decision Sciences*, *7*(1), 39–75. doi:10.4018/IJSDS.2016010103

Redmond, A., Smith, B., & Smith, D. (2014). The Curriculum Development of a BIM Resilience Program for the National Institute of Building Science Facility Module. *International Journal of 3-D Information Modeling, 3*(1), 49-60. doi:10.4018/ij3dim.2014010105

Rickards, T. (2016). Improving the Effectiveness of Research Supervision in STEM Education: Cloud-Based Multimedia Solutions. In L. Chao (Ed.), *Handbook of Research on Cloud-Based STEM Education for Improved Learning Outcomes* (pp. 343–356). Hershey, PA: IGI Global. doi:10.4018/978-1-4666-9924-3.ch021

Rickards, T., & Steele, A. (2016). Designing a Cloud-Based Assessment Model: A New Zealand Polytechnic Case Study. In L. Chao (Ed.), *Handbook of Research on Cloud-Based STEM Education for Improved Learning Outcomes* (pp. 357–376). Hershey, PA: IGI Global. doi:10.4018/978-1-4666-9924-3.ch022

Rizzardini, R. H., & Gütl, C. (2016). A Cloud-Based Learning Platform: STEM Learning Experiences with New Tools. In L. Chao (Ed.), *Handbook of Research on Cloud-Based STEM Education for Improved Learning Outcomes* (pp. 106–122). Hershey, PA: IGI Global. doi:10.4018/978-1-4666-9924-3.ch008

Ro, H. K., & McIntosh, K. (2016). Constructing Conducive Environment for Women of Color in Engineering Undergraduate Education. In U. Thomas & J. Drake (Eds.), *Critical Research on Sexism and Racism in STEM Fields* (pp. 23–48). Hershey, PA: IGI Global. doi:10.4018/978-1-5225-0174-9.ch002

Roberts, K. L. (2015). Interdisciplinary Course Development in Nanostructured Materials Science and Engineering. In H. Lim (Ed.), *Handbook of Research on Recent Developments in Materials Science and Corrosion Engineering Education* (pp. 212–229). Hershey, PA: IGI Global. doi:10.4018/978-1-4666-8183-5.ch011

Rogers, R. R., Winship, J., & Sun, Y. (2016). Systematic Support for STEM Pre-Service Teachers: An Authentic and Sustainable Four-Pillar Professional Development Model. In K. Dikilitaş (Ed.), *Innovative Professional Development Methods and Strategies for STEM Education* (pp. 73–90). Hershey, PA: IGI Global. doi:10.4018/978-1-4666-9471-2.ch005

Russo, M. R., & Brittain, K. (2014). Women and STEM. In V. Wang (Ed.), *Handbook of Research on Education and Technology in a Changing Society* (pp. 703–713). Hershey, PA: IGI Global. doi:10.4018/978-1-4666-6046-5.ch052

Sahlin, J. P., & Lobera, K. (2016). Cloud Computing as a Catalyst for Change in STEM Education. In L. Chao (Ed.), *Handbook of Research on Cloud-Based STEM Education for Improved Learning Outcomes* (pp. 12–30). Hershey, PA: IGI Global. doi:10.4018/978-1-4666-9924-3.ch002

Sakulich, A., El-Korchi, T., & Sisson, R. D. Jr. (2015). The Interdisciplinary, Project-Based Infrastructure Degradation Curriculum at Worcester Polytechnic Institute. In H. Lim (Ed.), *Handbook of Research on Recent Developments in Materials Science and Corrosion Engineering Education* (pp. 67–81). Hershey, PA: IGI Global. doi:10.4018/978-1-4666-8183-5.ch004

Sakulich, A., & Peterson, A. (2017). A Globally Focused, Experiential Educational System for STEM Fields: Measures for Intentionally Promoting Diversity. In M. Gray & K. Thomas (Eds.), *Strategies for Increasing Diversity in Engineering Majors and Careers* (pp. 176–200). Hershey, PA: IGI Global. doi:10.4018/978-1-5225-2212-6.ch009

Sciffer, S., & Shah, M. (2015). Widening the Participation of Disadvantaged Students in Engineering. *International Journal of Quality Assurance in Engineering and Technology Education*, *4*(1), 1–13. doi:10.4018/IJQAETE.2015010101

Seals, C., Horton, A., Berzina-Pitcher, I., & Mishra, P. (2017). A New Understanding of our Confusion: Insights from a Year-Long STEM Fellowship Program. In C. Martin & D. Polly (Eds.), *Handbook of Research on Teacher Education and Professional Development* (pp. 582–604). Hershey, PA: IGI Global. doi:10.4018/978-1-5225-1067-3.ch032

Sharpe, R. V. (2016). HBCUs: Efficiencies of Creating a Scientific Workforce Outta Fifteen Cents. In C. Prince & R. Ford (Eds.), *Administrative Challenges and Organizational Leadership in Historically Black Colleges and Universities* (pp. 199–212). Hershey, PA: IGI Global. doi:10.4018/978-1-5225-0311-8.ch011

Smith, L. K., Rooney-Varga, J. N., Gold, A. U., Oonk, D. J., & Morrison, D. (2016). Media Literacy as a Pathway to Bridge the Digital and STEM Divides: Interest Driven Media Projects for Teachers in the Trenches. In M. Urban & D. Falvo (Eds.), *Improving K-12 STEM Education Outcomes through Technological Integration* (pp. 23–43). Hershey, PA: IGI Global. doi:10.4018/978-1-4666-9616-7.ch002

Smith, P., Fulton, J., Irons, A., & Sanders, G. (2016). The Role of the Professional Doctorate in Developing Professional Practice in STEM Subjects. In K. Dikilitaş (Ed.), *Innovative Professional Development Methods and Strategies for STEM Education* (pp. 1–16). Hershey, PA: IGI Global. doi:10.4018/978-1-4666-9471-2.ch001

Smith-Walters, C., & Barker, H. L. (2015). Pre-Service Teachers' Self-Efficacy and Attitudes toward Learning and Teaching Science in a Content Course. In E. de Silva (Ed.), *Cases on Research-Based Teaching Methods in Science Education* (pp. 277–302). Hershey, PA: IGI Global. doi:10.4018/978-1-4666-6375-6.ch012

Srisawasdi, N. (2016). Motivating Inquiry-Based Learning Through a Combination of Physical and Virtual Computer-Based Laboratory Experiments in High School Science. In M. Urban & D. Falvo (Eds.), *Improving K-12 STEM Education Outcomes through Technological Integration* (pp. 108–134). Hershey, PA: IGI Global. doi:10.4018/978-1-4666-9616-7.ch006

Sung, W., Ahn, J., Kai, S. M., Choi, A., & Black, J. B. (2016). Incorporating Touch-Based Tablets into Classroom Activities: Fostering Children's Computational Thinking through iPad Integrated Instruction. In D. Mentor (Ed.), *Handbook of Research on Mobile Learning in Contemporary Classrooms* (pp. 378–406). Hershey, PA: IGI Global. doi:10.4018/978-1-5225-0251-7.ch019

Swaid, S. I. (2014). A Novel Strategy to Improve STEM Education: The E-Science Approach. In V. Wang (Ed.), *Handbook of Research on Education and Technology in a Changing Society* (pp. 860–871). Hershey, PA: IGI Global. doi:10.4018/978-1-4666-6046-5.ch064

Sweeny, R. W. (2017). The Politics of Video Games in STEM Education. In R. Shin (Ed.), *Convergence of Contemporary Art, Visual Culture, and Global Civic Engagement* (pp. 331–341). Hershey, PA: IGI Global. doi:10.4018/978-1-5225-1665-1.ch019

Taber, K. S. (2016). Constructivism in Education: Interpretations and Criticisms from Science Education. In E. Railean, G. Walker, A. Elçi, & L. Jackson (Eds.), *Handbook of Research on Applied Learning Theory and Design in Modern Education* (pp. 116–144). Hershey, PA: IGI Global. doi:10.4018/978-1-4666-9634-1.ch006

Taber, K. S., Ruthven, K., Howe, C., Mercer, N., Riga, F., Hofmann, R., & Luthman, S. (2015). Developing a Research-Informed Teaching Module for Learning about Electrical Circuits at Lower Secondary School Level: Supporting Personal Learning about Science and the Nature of Science. In E. de Silva (Ed.), *Cases on Research-Based Teaching Methods in Science Education* (pp. 122–156). Hershey, PA: IGI Global. doi:10.4018/978-1-4666-6375-6.ch006

Tang, K. C., Chiu, D. K., Chan, A. W., & Tang, J. K. (2016). Enterprise Content Management System for IT Training. *International Journal of Systems and Service-Oriented Engineering*, *6*(3), 32–54. doi:10.4018/IJSSOE.2016070103

Tang, M. S., & Karunanithi, A. T. (2016). A Media Interpretation of the Negative Impact of Cloud Technologies. In L. Chao (Ed.), *Handbook of Research on Cloud-Based STEM Education for Improved Learning Outcomes* (pp. 403–420). Hershey, PA: IGI Global. doi:10.4018/978-1-4666-9924-3.ch024

Tavernise, A., Bertacchini, F., Pantano, P. S., & Bilotta, E. (2015). Implementing a New Class-Lab: Guidelines for Integrating Innovative Devices in Pre-Service Teachers Practice. *International Journal of Digital Literacy and Digital Competence*, *6*(3), 33–49. doi:10.4018/IJDLDC.2015070103

Terrazas-Arellanes, F. E., Knox, C., Strycker, L. A., & Walden, E. D. (2017). Online Learning Tools for Middle School Science: Lessons Learned from a Design-Based Research Project. *International Journal of Information and Communication Technology Education*, *13*(1), 27–40. doi:10.4018/IJICTE.2017010103

Terrazas-Arellanes, F. E., Knox, C., & Walden, E. (2015). Pilot Study on the Feasibility and Indicator Effects of Collaborative Online Projects on Science Learning for English Learners. *International Journal of Information and Communication Technology Education*, *11*(4), 31–50. doi:10.4018/IJICTE.2015100103

Thien, N. D., Terracina, A., Iocchi, L., & Mecella, M. (2016). Robotic Teaching Assistance for the Tower of Hanoi Problem. *International Journal of Distance Education Technologies, 14*(1), 64–76. doi:10.4018/IJDET.2016010104

Thomas, U., & Drake, J. (2016). Grace under Fire: Examining the Underrepresentation and Persistence of Women in STEM Professions from an Ecological Systems Theory Perspective. In U. Thomas & J. Drake (Eds.), *Critical Research on Sexism and Racism in STEM Fields* (pp. 1–22). Hershey, PA: IGI Global. doi:10.4018/978-1-5225-0174-9.ch001

Ursyn, A. (2016). Teaching and Learning Science as a Visual Experience. In A. Ursyn (Ed.), *Knowledge Visualization and Visual Literacy in Science Education* (pp. 1–27). Hershey, PA: IGI Global. doi:10.4018/978-1-5225-0480-1.ch001

Van Sickle, M., & Koester, M. (2017). Musing on Unanswered Questions. In J. Bazler & M. Van Sickle (Eds.), *Cases on STEAM Education in Practice* (pp. 1–20). Hershey, PA: IGI Global. doi:10.4018/978-1-5225-2334-5.ch001

Vandergon, V. O., Reveles, J., Herr, N., Nguyen-Graf, D., Rivas, M., d'Alessio, M., & Foley, B. (2016). Engaging Students in Conducting Data Analysis: The Whole-Class Data Advantage. In L. Chao (Ed.), *Handbook of Research on Cloud-Based STEM Education for Improved Learning Outcomes* (pp. 172–197). Hershey, PA: IGI Global. doi:10.4018/978-1-4666-9924-3.ch012

Vebrianto, R., Rery, R. U., & Osman, K. (2016). BIOMIND Portal for Developing 21st Century Skills and Overcoming Students Misconception in Biology Subject. *International Journal of Distance Education Technologies, 14*(4), 55–67. doi:10.4018/IJDET.2016100105

Velázquez-Iturbide, J. Á., Debdi, O., & Paredes-Velasco, M. (2015). A Review of Teaching and Learning through Practice of Optimization Algorithms. In R. Queirós (Ed.), *Innovative Teaching Strategies and New Learning Paradigms in Computer Programming* (pp. 65–87). Hershey, PA: IGI Global. doi:10.4018/978-1-4666-7304-5.ch004

Webb, A. M., Knight, S. L., Ben Wu, X., & Schielack, J. F. (2014). Teaching Science with Web-Based Inquiry Projects: An Exploratory Investigation. *International Journal of Virtual and Personal Learning Environments, 5*(2), 57–68. doi:10.4018/ijvple.2014040105

Weber, A. S. (2016). Strategic Planning for Cloud Computing Adoption in STEM Education: Finding Best Practice Solutions. In L. Chao (Ed.), *Handbook of Research on Cloud-Based STEM Education for Improved Learning Outcomes* (pp. 1–11). Hershey, PA: IGI Global. doi:10.4018/978-1-4666-9924-3.ch001

Williams, P. H., & Lee, C. K. (2017). Collaborations between Elementary Schools and Higher Education: Lessons Learned from Community-Engaged Science Education Efforts. In C. Crosby & F. Brockmeier (Eds.), *Community Engagement Program Implementation and Teacher Preparation for 21st Century Education* (pp. 82–104). Hershey, PA: IGI Global. doi:10.4018/978-1-5225-0871-7.ch005

Xia, B. S. (2017). A Pedagogical Review of Programming Education Research: What Have We Learned. *International Journal of Online Pedagogy and Course Design, 7*(1), 33–42. doi:10.4018/IJOPCD.2017010103

Yantz, J., Smith, B. D., Rowell, G. H., Cheatham, T., Nelson, D., Stephens, D. C., & Tenpenny, E. B. (2015). Using Multidisciplinary Research Experiences to Enhance STEM Learning through Undergraduate, Team-Based, Summer Research Projects for At-Risk Students. In E. de Silva (Ed.), *Cases on Research-Based Teaching Methods in Science Education* (pp. 198–220). Hershey, PA: IGI Global. doi:10.4018/978-1-4666-6375-6.ch008

Yu, L. (2016). Experience Teaching Introductory Programming Courses. *International Journal of Quality Assurance in Engineering and Technology Education, 5*(1), 40–53. doi:10.4018/IJQAETE.2016010104

Zaharias, P., Chatzeparaskevaidou, I., & Karaoli, F. (2017). Learning Geography Through Serious Games: The Effects of 2-Dimensional and 3-Dimensional Games on Learning Effectiveness, Motivation to Learn and User Experience. *International Journal of Gaming and Computer-Mediated Simulations, 9*(1), 28–44. doi:10.4018/IJGCMS.2017010102

Index

I

J

K

L

M

N

O

P

R

S

T

U

V

Z

www.ingramcontent.com/pod-product-compliance
Ingram Content Group UK Ltd.
Pitfield, Milton Keynes, MK11 3LW, UK
UKHW050945290926
15470UKWH00005B/1639

9 781522 521846